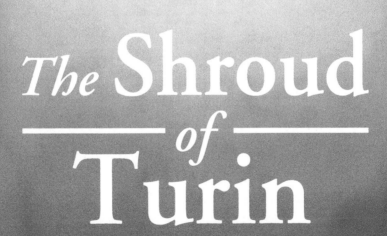

The Shroud
—— of ——
Turin

The Shroud

of

Turin

FIRST CENTURY AFTER CHRIST!

Giulio Fanti

Pierandrea Malfi

With an in-depth study by Marco Conca

PAN STANFORD PUBLISHING

Published by

Pan Stanford Publishing Pte. Ltd.
Penthouse Level, Suntec Tower 3
8 Temasek Boulevard
Singapore 038988

Email: editorial@panstanford.com
Web: www.panstanford.com

British Library Cataloguing-in-Publication Data
A catalogue record for this book is available from the British Library.

The Shroud of Turin: First Century after Christ!

ISBN 978-981-4669-12-2 (Hardcover)
ISBN 978-981-4669-13-9 (eBook)

Printed in the USA

Contents

Foreword

The use of the terms "first author" and "second author" in the book might come to the reader as a surprise. The reason is that two authors with different research paths and opinions about religion have collaborated constructively on the subject of the Shroud. The numerous discussions held together have both advanced them personally and improved the book's content. Clearly, outside a strictly scientific context, religious considerations cannot merely be the outcome of an honorable compromise between authors. Whenever compromise is impossible, the opinions of each author are therefore distinguished. For the sake of simplicity, we will refer to them as first author and second author.

Preface

The Shroud of Turin, the linen sheet that according to tradition would have enveloped the body of Jesus in the sepulchre, is still the center of interest of public opinion, mass media, and the scientific world. Less than one year after the publication of his last work, written together with Saverio Gaeta, Professor Giulio Fanti, from the University of Padua, comes with a new book, this time written in cooperation with Pierandrea Malfi.

I am glad to be asked to write the preface to this further work of Professor Fanti, since I had the chance to know him also beyond his research.

One may ask, What is the point of publishing a new book about the Shroud? As it is well known, the scientific research on the Turin relic has been officially closed in 1988, the year of the radiocarbon test that decreed the controversial medieval dating. What news can still emerge? Actually, even if research is officially closed, many scholars are continuing their activities, carrying out increasingly accurate investigations, and putting forward new interesting hypotheses. In 2002 the so-called Shroud Science Group was formed. It is composed of more than 100 researchers of different nationalities, most of who are American, and is coordinated by Professor Fanti. The group discusses on the web the most interesting news about the Holy Linen: it already organized an international conference on the Shroud in Dallas in 2008 and another one, in the United States as well, was scheduled for 2014.

Further than this, in the last decade around 20 articles describing new hypotheses and scientific acquisitions on the Shroud have been published in specialized international reviews. It does not appear that other relics, considered as such by the Catholic religion or other confessions, ever had the honor of such consideration in specialized reviews.

This is an overview that well paints a picture of how the Shroud is important also from the scientific point of view, despite the closing of official investigations in 1988.

How is it possible that the studies go on if the official research is closed, no analyses in situ are allowed, and no authorization for sample taking is granted? One has to be reminded that modern scientific tests do not need to handle samples of big dimensions and it is not always necessary to have the object of the research on hand (though it would be obviously more advisable). The number of photographs, from those taken in the past years to the recent ultrahigh-resolution ones taken by the Italian company Haltadefinizione, together with the samples officially taken in the past and now in possession of the various scholars throughout the world are sufficient, according to these researchers, for carrying on new analyses.

It is well known that the same Professor Fanti came into possession of some samples collected during the famous examinations carried out in 1978, the year in which also the Shroud of Turin Research Project (STURP) and the Italian group led by Giovanni Riggi di Numana and Pierluigi Baima Bollone took other samples. These consist of a few threads and linen fibers collected from different areas of the Shroud, but they are, according to the first author, more than enough for the microscopic tests needed to unveil the mysteries still concealed in the holy cloth.

Starting from the analyses of the chemical and mechanical properties of some Shroud samples, Professor Fanti, in a research project of the University of Padua, in cooperation with other professors of the University of Bologna, Modena, Parma, and Udine, sustains to have demonstrated that the radiocarbon dating test carried out on the Shroud in 1988 is unreliable from a statistical point of view. Therefore, results would be scientifically meaningless.

So, what are the innovations described in this book? Besides the as much understandable as possible description of the alternative linen dating methods, one of the most interesting chapters refers to an innovative numismatic investigation that would lead to further Shroud dating to the first millennium A.D.

The author's analysis involved the research and study of very rare coins minted during the Byzantine Empire and depicting the face of Christ. These faces have been compared to that of the Man of the Shroud. The surprising result, considering the number of figurative matches, was that the engravers should have been taken the Shroud face as a sample for reproducing the face of Christ during the centuries of the Byzantine coinage.

On the basis of this probabilistic counting, a theme highly supported by Professor Fanti, authors affirm that the Byzantine engravers would have seven chances in one billion of billions to hit all the particular features identified on the faces portrayed on the coins without having seen the Shroud. It is like saying that, referring to the roulette, it would be much easier to hit for 10 consecutive times the number 36 rather than an engraver having depicted the face displayed on the coins without having seen the Shroud.

Some typical details of the Shroud reproduced on the coins, such as the skewed nose, the eyes shut, asymmetric length of the hair, longer on the right side, closed eyes, and a long nose, are hints of the reference to the Shroud model. These particulars are according to the authors unthinkable to be reproduced by an artist who wanted to depict "the most handsome of the sons of men" (Psalm 45:2) instead of the tortured Jesus of the Shroud.

Also, pages that go through the different hypotheses of the body image formation do not miss, with an explanation on the most reliable possibility, that, according to Professor Fanti, it was an intense electric charge that, through the so-called corona discharge, could have reproduced many image peculiarities. Even if it was like that, the question about what would have triggered this high electric charge still remains.

Authors sustain, therefore, that radiocarbon dating is not reliable and that the Shroud is dating back to the first century A.D., accordingly compatible with the age in which Jesus lived. It is also known that the Man of the Shroud shows wounds and blows perfectly matching with the signs of the Passion of Christ described in the Gospels.

A good Christian knows that God's method leaves Man's freedom always untouched. Therefore, it would be a serious mistake to think of "demonstrating" something with the Shroud—least of all, demonstrating the basis of Christianity, the Resurrection of Jesus. It is true that until the present days no Gospel's verse has been contradicted by scientific discoveries; rather, as the historical, archeological, and epigraphic studies made progress in the last decades, new confirmations about the historicity of the Gospels came. But it is likely that even in the future, the same light will be shining for people who want to believe and will be too faint for people who do not want to believe, this particularly for artifacts like the Shroud of Turin.

At the same time, it is positive that there are scientists and researchers willing to face all the questions remaining open. Science goes on for trials, hypotheses, discussions, and debates.

For example, the increase of the doubts of medieval dating is noteworthy. Asking questions, formulating hypotheses, trying to give answers, . . . we always will need it. Here is why any contribution based on honest research is welcome.

Andrea Tornielli

Acknowledgments

We gratefully thank the Holy Father Francis for the great interest shown towards the Italian version of this book.

Many thanks to all those who have contributed to this book. We gratefully acknowledge the following persons who collaborated in the translation: Giulia Azzari, Tommaso Caniato, Sara Guth, Federica Paglialunga, and Ira Sohn.

PART I

DESCRIPTION AND TRACES OF THE SHEET THAT CHALLENGES SCIENCE

Chapter 1

The Shroud: An Identikit

Almost everyone, at least once, has heard of the so-called Holy Shroud, the Shroud of Turin, or more simply the Shroud, perhaps referring to the fact that the body image depicted on the cloth had been painted by a medieval artist. However, very few people had the chance to investigate the number of mysteries that make the Shroud a unique object in the world, not only impossible to reproduce but also impossible to be explained only by scientific means. In this first chapter, we are going to conduct readers into these mysteries and, in the end, lead them toward some explanatory hypotheses.

1.1 The Subject of the Investigation

The Shroud is the most studied object in the world, both in the archaeological and obviously in the religious field. On this subject, hundreds of books in many different languages and dozens of scientific papers have been written, and articles and notes published every day in newspapers and on the web are countless.

The Shroud of Turin: First Century after Christ!
Giulio Fanti and Pierandrea Malfi
Copyright © 2015 Pan Stanford Publishing Pte. Ltd.
ISBN 978-981-4669-12-2 (Hardcover), 978-981-4669-13-9 (eBook)
www.panstanford.com

The first book on the Shroud containing scientific data was written by Alfonso Paleotto [107] in 1598, a few years after that the Shroud had been moved to Italy, in Turin, in 1578. This data unequivocally demonstrates the long-lasting interest in the Holy Shroud shown all over the world not just by believers.

The Catholic Church does not impose devotion toward this Relic, even if the tradition states that it was the Jesus' burial shroud after his death by crucifixion, and the several scientific analyses performed have not been able to question this tradition.

In fact, the double body image imprinted on the sheet is neither explainable nor reproducible, which means that nobody till this day has been able to implement an identical copy both at the macroscopic and at the microscopic level.

In the past, the Shroud has been popularly called the *Holy Shroud*, a term now used essentially in the religious field; scientists named it the *Shroud of Turin* and here we simply call it the *Shroud*.

It is a linen cloth that for sure enveloped a man, the so-called *Man of the Shroud*, whose identity is still unknown. Lots of evidence supports tradition and many features match with the Biblical description of Jesus, but currently any conclusion regarding the identity of the Man of the Shroud it is just up to one's interior consciousness, since science does not have the "case solution" or any indisputable evidence yet and yet the Man depicted on the linen sheet does not have a name.

Someone could say that this is exactly the proof of its divine origin: if the Man of the Shroud had a name, God would impose His will over mankind's free will. Comparing the parts of the canonical Gospels, describing also the Roman punishment of flagellation inflicted on Jesus as the prelude of death sentence execution by crucifixion,[1] with the elements acquired on the Shroud, several are the clues leading to the identification of the Man of the Shroud with

---

[1]Usually scourging was a separate punishment that did not consider a second condemnation like crucifixion. In the specific case of Jesus, Gospels tell that Pontius Pilate, the prefect of the Roman province of Judaea, after having condemned Jesus to scourging, was pushed by Jewish people to crucify him. The fact that the Man of the Shroud shows unusual traces of both punishments is strong evidence in favor of the hypothesis that identify the Man of the Shroud with Jesus Christ. For details on the crucifixion technique of Man of the Shroud see Section 10.5.

Jesus of Nazareth, but everyone can reach a personal conclusion without any interference with one's own belief.

The Italian term *Sindone* (Shroud), from the Ancient Greek *sindón* (σινδών), means a *sheet* or *piece of cloth for a specific use*. In turn, it seems that the term sindón derives from sindia or sindien, which is an Indian cloth.

Onto the Shroud we can observe different signs, more or less important, and not easily distinguishable at first glance, even because their partial superposition makes the identification quite complicated. We can see the following:

- A double specular body image,[2] front and back, of a man corpse (Fig. 1.1)
- Blood stains[3] in correspondence with the wounds of the man enveloped into the cloth (Fig. 1.2)
- Water stains (Fig. 1.3)
- Burn holes caused by the A.D. 1532 Chambéry fire,[4] by other fires, or accidents (Fig. 1.4)

Until June 2002, before the heavy intervention performed, it was possible to observe the patches sewn by the Poor Clare nuns in 1534 to partially repair the damages caused by the fire two years earlier. Now these patches have been removed and holes due to fire have been left exposed [71].

The faint yellow linen sheet is almost rectangular, compact, and robust. The color of body image is darker but not so much contrasted compared to the background because of the yellowing of the sheet due to aging.

---

[2] Since the Shroud enveloped a man, the imprinted image is specular. The right arm in Fig. 1.2 is actually the left arm of the Man of the Shroud.

[3] The red stains have been independently identified as human blood by A. Adler and P. Paima Bollone. The first author Giulio Fanti discovered blood particles among the dusts vacuumed from the Shroud.

[4] There are traces of accidents happened before the Chambéry fire; in fact, in a copy of the Shroud of 1516 ascribed to Dürer there are no signs of the fire, but some details, like four sets of L-shaped holes named Poker Holes are present. Shroud history during the centuries is rather complicated. Chapter 2 will briefly describe it.

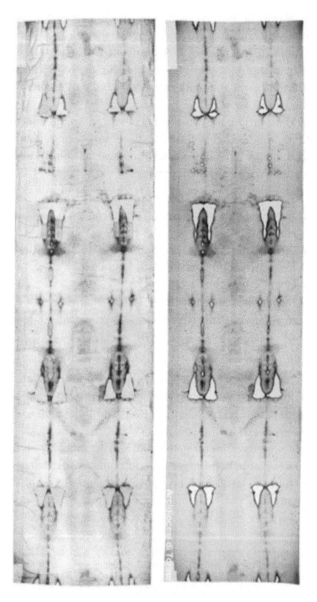

Figure 1.1 The Shroud before 2002 intervention (on the left, A. Guerreschi) and after (on the right, Archdiocese of Turin).

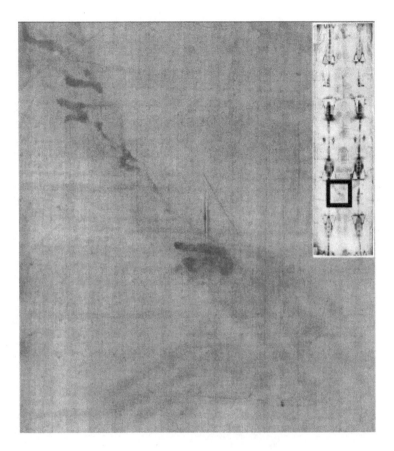

Figure 1.2 Blood stains in correspondence with the wrist and the left arm. In the frame on the top-right corner is shown the position of the detail on the cloth.

No direct weight measurements have been taken on the relic, but indirect tests carried out also by Fanti on small parts of the cloth lead to a total weight of 1.05 kg (2.31 lb), obtaining a mass per unit area of 0.22 kg/m^2 (0.045 lb/ft.^{2}).

The cloth is about 0.34 mm (0.013 in.) in thickness and the diameter of linen yarns is 0.25 mm (0.0098 in.) on average. The traditional dimensions were quoted as 436 cm (14.30 ft.) of length and 110 cm (3.61 ft.) of width. After the 2002 intervention [16], the

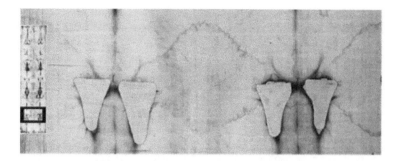

Figure 1.3 Example of 3 water stains. In the picture are also evident 4 patches sewn onto the cloth by Poor Clare nuns to repair the damages caused by the Chambéry fire.

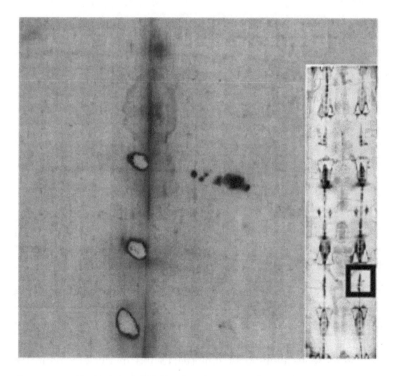

Figure 1.4 Example of holes preceding the Chambéry fire and some little stains probably caused by the water used for extinguishing the fire. According to some researchers, it is more likely that holes would have been caused by a corrosive substance than by a fire.

new official measures [71] have increased by some centimeters[5] also because patches and some folds have been removed.

Sewn onto one of the long sides of the Shroud is the so-called "side strip," a 8.5 cm (3.35 in.) wide strip of linen incomplete at its ends. Textile features of the side strip correspond to the main cloth to which it is attached. Probably the side strip was originally one piece with the cloth and it was cut apart and reattached in the present position so that the Shroud's image becomes centered and balanced on the cloth.

The seam between the strip and the rest of the cloth is very peculiar and typical of textiles used in the Middle East in the first millennium A.D. The manufacture of the Shroud is by contrary rudimental and the linen handspun.

The extraction of the fibers from the stem of the flax plant and the in-depth analysis of the microscopic characteristica of the fibers will be discussed in Chapter 5. Here, we can point that every yarn is composed of an average of about 200 fibers[6] and has a diameter of a quarter of millimeter (0.0098 in.), variable from yarn to yarn. Every fiber has an average diameter of 0.013 mm (0.00051 in.),[7] but it is also possible to find fibers with a diameter superior to 0.020 mm (0.00079 in.) or inferior to 0.007 mm (0.00027 in.) [15, 118,

---

[5]From the work of G. Ghiberti [71] a detailed comparison with the 2000 measurements: *At the end of the operations executed on the cloth, in order to acquire the definitive measurements of the Shroud as it would have been put back on its case in the Chapel, a verification was performed. It was carried out by Bruno Barberis (director of the International Centre of Sindonology in Turin (A/N)) and Gian Maria Zaccone (scientific director of the Shroud Museum (A/N)). They established an increase of some centimeters in comparison with the measurements taken in 2000. Considering horizontal the body image, with the frontal side on the left and the dorsal side on the right, the lower side measured 437.7 cm (14.360 ft.) in 2000 and 441.5 cm (14.485 ft.) in 2002; the opposite side (less substantial since the extremities are constituted only by the Linen; in fact, in the past, two portions of the original cloth had been removed) measured 434.5 cm (14.551 ft.) in 2000 and 442.5 cm (14.518 ft.) in 2002; its height (with a relative value for the above-mentioned reason) of 112.5 cm (3.691 ft.) and 113 cm (3.707 ft.), respectively, on the left and on the right in 2000, was in 2002 113 cm (3.707 ft.) on the left and 113.7 cm (3.730 ft.) on the right."*

[6]Fanti counted 188±5 fibers in a thread, group discussion named Shroud Science 31/7/2007.

[7]In order to simplify the comprehension of numerical values for the readers, in the text we refer only to the submultiple millimeter (mm) of the meter, not using micrometers (thousandths of a millimeter).

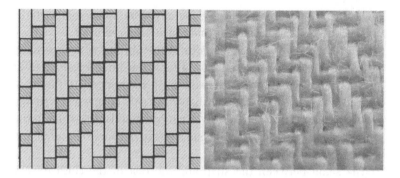

Figure 1.5 Example of the three-over-one weave pattern: the model on the left, the example on textile on the right. Photo's source [56].

119, 144]. As a basis for comparison, a human hair has an average diameter of 0.080 mm (0.0031 in.).

As Fig. 5.6 in Chapter 5 shows, Shroud yarns are "Z" twist,[8] meaning the spindle was rotated clockwise, opposed to the more common S-twist spun. This element recalls the fact that this kind of fabric was in the past used for dressing High Priests, exactly as described in the Exodus, in which for 21 times is mentioned the expression (translating from Latin) *bysso retorta*. The expression *bysso retorta* is translated in the English version of the Exodus as "fine linen twined." However, this does not properly convey the real meaning of *bysso retorta*, since the linen is not simply "twined" (as an S-twist spun would be), but it is countertwined. Therefore, the Book indicates the "Z" twist and not the classic "S" twist.

The irregular fabric weaving could suggest that a primitive loom was used to trim the Shroud; in fact there are faults in the weave and errors. The weave pattern is diagonal and three-over-one, obtained by the weft passing over three warp threads, under one, over three, and so forth (Fig. 1.5).[9] The twill, which gives to the cloth the herringbone aspect, forms "stripes" wide, about 11 mm (0.43 in.), and it is another element indicating that the cloth was extremely refined and precious for that period.

---

[8]The "Z" twist is extremely rare.
[9]See Section A.7 and Fig. A.5 of the appendix for further information about the 3–1 weaving technique.

1.2 Current Location and Conservation

At the time of the Chambéry fire in 1532, the Holy Linen, hedged in a reliquary covered with silver, was folded into 32 layers. Previously, the Shroud came under an incident that left four sets of holes: the sheet should have been folded into four when this damage happened.

Furthermore, other watermarks caused by water absorption can be identified. Presumably, in remote times the cloth was conserved inside an earthenware jar [75, 76].

The Shroud was moved to Turin in 1578 and the architect Guarino Guarini (1624–1683) was entitled to the construction of a magnificent chapel to preserve it. The Linen was finally transferred in 1694 and for almost three centuries it remained rolled up in a reliquary placed on the altar.

That reliquary did not protect the cloth from temperature and humidity variations, and the Shroud was also exposed to the atmospheric contamination of the industrial city of Turin.

In the casket, which was an optimal context for the development of bacteria, the environment deteriorated the fabric. The Linen was discovered to be infested by microbes, bacteria, and mold.

Additionally, rolling the Shroud on a cylinder caused wrinkles that damaged the Linen and the blood stains' integrity.

On September 7, 1992, Cardinal G. Saldarini organized an inspection of the Shroud, together with a qualified international commission composed of experts in different disciplines.

The conclusion of the commission was the following [63, p. 217]:

(a) The Shroud would have been kept in an extended, plain, and horizontal position.

(b) The Shroud would have been kept in a reliquary with bulletproof and waterproof glass, in the absence of air and in the presence of an inert gas. It would have been protected from light and maintained in constant climatic conditions.

(c) The reliquary would have been used also during the public exhibitions, with the purpose of avoiding any handling of the Shroud.

Such conditions obviously imposed a radically different way of preservation from that of rolling up the Shroud on a cylinder and, above all, the need to build a proper reliquary. Therefore, the second phase of the operation started.

On April 17, 1998, the Shroud was inserted, in a horizontal position, in its new case that suited the requirements for public exhibitions, realized by the "workshop" Bodini. The relic was monitored in order to be kept in constant climatic conditions: inside the airtight case there was a mixture of argon (an easily available inert gas that prevents the development of microorganisms) and water vapor [63]; the quantity of oxygen was continuously controlled[10] in order not to exceed 0.01%–0.1%. The temperature was also controlled at around 18°C (64.4°F).

A new case was built in 2000 by the Italian company Alenia Spazio for the standard conservation of the Shroud out of public displays. The requirements were as follows [63]:

(a) The Shroud would have been stored fixed and flat, 4.5 m (14.76 ft.) × 1.3 m (4.27 ft.), on a portable support sliding on runners.

(b) The reliquary would have been placed with a highly airtight system in order to avoid the continuous monitoring of the internal atmosphere (humid argon mixed with 0.2/0.3% oxygen), the system used in the previous case.[11]

(c) Internal overpressure to avoid external environment contamination.

(d) Internal pressure and temperature constant control.

(e) The reliquary would have been built of material free of outgassing phenomena (volatile substance release).

(f) The Shroud would have been observed, whenever needed, without taking it off from the inert atmosphere in which it was conserved.

---

[10] According to some information received by the first author on 13/12/2007, it resulted that environmental conditions were actually changed: an argon atmosphere containing about 10% oxygen and 50% relative humidity. These conditions seemed more suitable for the relic, since 10% oxygen contrasted the proliferation of anaerobic bacteria, which live in the absence of air, and limited the development of other bacteria.

[11] According to the same information of 2007 (see footnote 10 on p. 12), the oxygen percentage did not change.

(g) Possibility of taking the Shroud out quickly in the case of emergency.

(h) The reliquary should have been neither too heavy nor big in order to facilitate its handling in the chapel in which it was alloyed.

The crystal lid, 29 mm (1.14 in.) thick and bulletproof, allows the complete seeing of the Shroud. The sheet was placed in the new case on December 22, 2000.

Every archaeological find has to be submitted to conservative interventions so that time does not deteriorate it. In the specific case of the Shroud, it has been necessary to study and solve the following three problems related to its conservation.

Fabric Conservation

It is not a problem in itself, since linen is resistant during time if not exposed to particular adverse environmental conditions. One of the problems of the conservation was the rolling or the folding of the cloth. According to some American experts, A. Adler and L. Schwalbe, the Shroud repository should not be made of steel. An inert material like glass is definitely more effective, since radiations, also those caused by cosmic emissions, ionizing the metallic materials of the case, destroy linen integrity. To reduce this kind of damage, it would be appropriate to keep the Shroud in a vertical position and keep any kind of plastic material away from it; furthermore, the sheet should have been placed in a deep underground facility. The environmental temperature has to be maintained relatively low but not under certain values, otherwise linen fibers could stiffen and break. Finally, protection against catastrophic events like earthquakes or floods or terrorist attacks must be considered.

Body Image Conservation

Originally, the linen color should have been lighter, but passage of time made it darker. The process of yellowing of linen derives from a chemical alteration of the fibers, these last incurring into dehydration exactly as the superficial cover of the body image does. Basically, this image corresponds to premature aging located on the linen fiber cover. By consequence, the body image background is getting yellower and yellower, with the risk that in a long period it will become of the same color of the body image itself. Then, the contrast between the image and the background will disappear and the observer will not be able to see the image anymore. To avoid this, it is of extreme importance to reduce the aging process. This depends on various parameters, such as temperature, light, and atmosphere. As far as temperature is concerned, it would be recommendable to keep the relic at very low temperatures, between −150°C (−238 °F) and −200°C (−328°F), because at these temperatures every chemical process, including aging, is slowed down. But this goes against the guidelines on fabric conservation, so it will be necessary to reach a compromise.

As for light, it is generally recognized that it accelerates fiber aging and, particularly, the aging of the cellulose it contains.[12] Ultraviolet (UV) rays with a lower wavelength are primarily responsible of this kind of process, and sunlight is also composed of these rays.

For the Shroud to be conserved at best, it has to be placed in a light-tight recovery. In fact, during the standard conservation in its case, it has a protection, since the reliquary is covered by an opaque plate against light radiations. However, during public exhibitions or conservative interventions, the relic is obviously exposed to solar rays.

Blood Stain Conservation

Blood stains are currently in the form of little crusts very sensitive to any fabric deformation. During the rolling and unrolling operations, many crusts were broken and distributed alongside the cloth. Now that the Shroud is kept flat and horizontal this problem is partially solved. Even light, especially UV rays, can modify blood color and the DNA contained in it. This is another reason why the relic has to be protected from light.

With these requirements satisfied, after the 2010 public display, Pope Francis authorized a new exhibition in 2015 (see the in-depth analysis in Section A.1 of the appendix).

---

[12] Some researchers gave a demonstration of the effect of light on the linen cloth, hypothesizing as primary cause of body image formation the emission of light during the Resurrection.

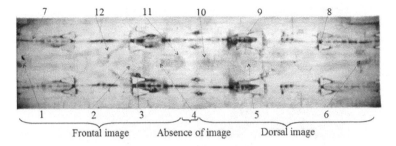

Figure 1.6 Negative body image and visible signs on the Shroud. (1) Scourge wounds on the legs. (2) Water stains caused by an accident antecedent to 1532. (3) Chest wound. (4) Fabric creases. (5) Scourge hits on the back. (6) Nail injury to right foot. (7) Charred lines due to the Chambéry fire in 1532. (8) Patches sewn by the Poor Clare nuns after the Chambéry fire. (9) Bruises perhaps derived from body rubbing on the Cross. (10) Head wounds caused by the crown of thorns. (11) "ϵ" or reversed "3"-shaped forehead wound. (12) Nail injury to the left wrist.

1.3 The Body Image: A Description

The most interesting feature of the Shroud consists of the two front and back images of a full-size lying human body separated by a space that does not bear body traces (Fig. 1.6).

The image is much clearer in the black-and-white negative than in its natural sepia color because gray scale is inverted: the eye–brain system perceives light areas as more protruding than the darker ones, but in the body image the darker areas are more prominent, as, for example, the nose tip (Fig. 1.7). This is the reason why image body features are more visible in the black-and-white negative.

Since the image lacks a sharp outline, the observers can see it from at least a 1 m distance. The closer they come, the more the image disappears from their visual perception.

According to the specialists who studied it, the body on the sheet corresponds to an about 30-year-old man with a beard and long hair. The body is well proportioned and muscular and it seems accustomed to physical labor [20, 145].

The Man of the Shroud had been enveloped in the sheet (Fig. 1.8) after his death, as rigor mortis shows [62], together with blood and serum runoff from the chest wound: in fact, if blood comes out

Figure 1.7 Positive image (above) and negative image (below) of the Pantocrator Icon in the St. Catherine's Monastery in Sinai (on the left) and the face of the Man of the Shroud (on the right). It is clear how the icon image loses any relevance in the negative version, whereas for the human eye the negative image of the Shroud face is easier to capture. To enhance the comparison with the related positive image, in this illustration, the Shroud face has not been reproduced overturned from left to right as it is done traditionally.

separated it means that there is no internal flow of blood and the heart has stopped.

The corpse was laid out on half of the Shroud, the other half then being drawn over the head to cover the body. Two imprints were left, one dorsal and one frontal (Fig. 1.9).

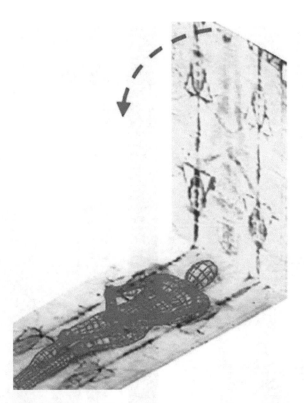

Figure 1.8 Computer representation of the position of the Man of the Shroud and of his enveloping in the sheet.

The frontal image is 195 cm (6.398 ft.) long; instead the dorsal one is 202 cm (6.627 ft.) (referring to the measurements taken from the pictures made by Giuseppe Enrie [48] in 1931); from these dimensions it is not easy to estimate the height of the Man of the Shroud, because after rigor mortis had developed on the Cross, his head was bent downward and feet extended. Further, it was likely that many joints had been dislocated due to the operations related to crucifixion.

By means of a computerized anthropometric analysis [56], it has been verified that the frontal and dorsal imprints of the Man of the Shroud are anatomically superimposable on a manikin representing him and it was shown that the Man's racial features can be attributable to the Semitic race.

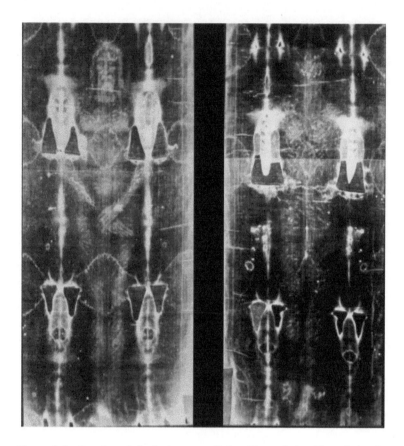

Figure 1.9 Front and back images of the Man of the Shroud in the photographic negative (G. Enrie).

If anthropometric indices of the Man of the Shroud match with those of a man, the same cannot be stated about those of pictorial images (e.g., the copy attributed to Dürer); this demonstrates that in the ancient times anatomic knowledge and the techniques of drawing were lacking in reproducing human details with exact proportions.

His shoulders seem lifted up like if the corpse would have been laid down on a trough or a soft support, probably containing substances for burial, like mineral salts and aromas, and spices [62].

Figure 1.10 Numerical manikin resulting from the compatibility analysis between frontal and dorsal images of the Shroud. For the values of the angles see footnote 13.

His head appears flexed downward[13] (Fig. 1.10) and maintained in that position by rigor mortis: in fact, on the frontal imprint there is no evidence of the neck, but it results extended in the dorsal image. Also, there is no evidence of the nape on the Linen and this confirms the onward position of the head due to cadaverous rigidity. Forearms and hands crossed over the pelvis, the left one on the right one, are clearly visible. Hand fingers are elongated, probably because the image has been distorted by the sheet enveloping; thumbs are not visible because, according to P. Barbet's [20] opinion, the median nerve was injured by the nail on the wrists or, according to more recent studies, because of the entrainment of the *flexor pollicis longus* tendons, while the nails were driven through the wrists [23].

The left foot imprint is less distinct than the right; over the sole of the right foot in fact a heel bone and fingers are evident. The right calf imprint is much more complete than the left, indicating that during the development of rigor mortis the left leg was slightly flexed, appearing shorter by consequence. The right foot was placed on the Cross and the left was placed on the instep of the right. Probably the feet were impaled together with a single nail.

1.4 Body Image: Typical Features

The double, front and back, body image of the Man of the Shroud reveals such peculiar characteristics that, until now, modern sciences could not reproduce all together at one time on a single

---

[13]The angles in Fig. 1.10 are as follows: angle α of the head $30° \pm 4°$, the angles β e γ of the right and of the left leg $19.5° \pm 3°$ and $23.5° \pm 3°$, respectively, and the angle δ of the right foot $34° \pm 2°$ and of the left foot $30° \pm 2°$ [62].

cloth. Currently it is therefore impossible to explain how the Shroud image has been created. Being considered a relic, it is understandable that someone would talk about a miracle referring to the formation of the image, but obviously, by the side of the science, this justification cannot be reasonable. From the scientific point of view, the study of the sheet led to several formulations of hypotheses that try to produce quite reliable, even though not completely satisfactory, explanations.

Before starting the analysis of these hypotheses, it is necessary to know at least some of the chemical and physical properties that make the Shroud a unique study case in the world.

The image is a superficial phenomenon at the level of the linen fabric. Using a brief scientific language, the perception of the Shroud image can be explained in this way: the variations on the cloth structure level involve variations of the incident light reflected by the sheet, which shape the image perceived by our eye–brain system. In other words, when we observe the Shroud, the light rays that hit an area of the cloth come out with different characteristics from those they came in from point to point, because the Linen surface is not identical in all its areas at a microscopic level. According to the area involved, the reflected light comes out more or less intense and with different chromatic characteristics, exactly as solar light that, hitting a polychromatic mosaic tile, comes out bringing the chromatic information of each tile. Once the rays reach our eyes, the retina transforms the light information in electric impulses that are elaborated as an image by our brain. In fact, light is captured by the eyes, but indeed, we see with our brain.

In 2005, 24 scientists of the Shroud Science Group[14] described 187 peculiar characteristics [64] of the Shroud image. We reported the paper text in Section A.2 of the appendix.[15] Here we will focus on some of the most important features, first from the chemical, physical, and optical points of view and then reporting some general notes.

---

[14] Constituted in 2002 and coordinated by Fanti, the Shroud Science Group is formed by about 100 scientists, mostly American (there are professors, researchers, and scholars). For further details the reader can refer to the end of Chapter 2.

[15] For the paper's bibliography the reader can download the work from http://www.shroud.com/pdfs/doclist.pdf.

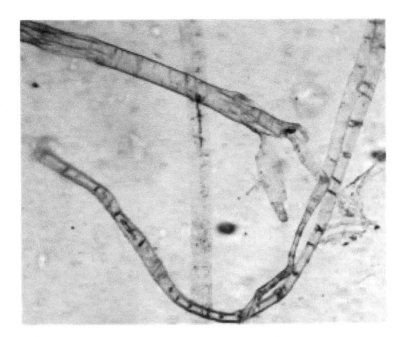

Figure 1.11 Colored fibers of the body image (darker) compared to one uncolored fiber (in the center) coming from the STURP-1EB sample given by R. Rogers to the first author.

From the chemical point of view, the image is due to a molecular modification of the surface of the linen fiber[16] constituted of polysaccharides (chains of glucose[17]). These polysaccharides underwent an alteration as a consequence of an acting-at-a-distance phenomenon (Fig. 1.11). In particular, the chemical reaction consists of dehydration with oxidation and conjugation (acid–base reaction). The image is not composed of painting pigments or other substances of that kind. The minimal evidence of pigments found among the vacuumed particles[18] of the Shroud is not even sufficient for reproducing a hundredth part of the whole image. Probably these

---

[16]A linen thread of the Shroud is on average composed of about 200 fibers with a diameter that can vary from 0.020 mm (0.00079 in.) to 0.007 mm (0.00027 in.). The polysaccharides that constitute every single linen fiber and their layer organization are described in detail in Chapter 5 and in the appendix.

[17]See footnote 7 on p. 168.

[18]For further details see Chapter 5 and Chapter 9.

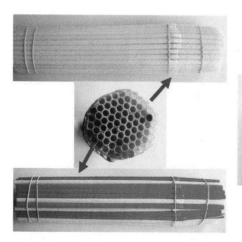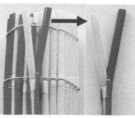

Figure 1.12 On the left, the first superficiality level is shown by the Shroud linen thread model, magnified 300 times, constituted of drinking straws; on the right, the second level of superficiality is highlighted by the fact that, removed from the colored layer, the straw (fiber) is uncolored.

traces derive from later contaminations, when artists were allowed to physically touch the Shroud with their paintings in order to create another relic. In support to this hypothesis is the fact that in the vacuumed particles have been found red particles (iron oxide) and particles of other colors (e.g., blue lapis lazuli).

From a physical point of view, the body image has two levels of superficiality. The first level consists of the fact that the image resides on the outermost layer of the linen fibers and the image goes just two or three fibers deep into the thread (Fig. 1.12). The superficial image then disappears if a colored thread goes under another thread.

A second level of superficiality consists of the fact that the coloration of every fiber constituting the image is only superficial: the polysaccharide cover, approximately 0.2 thousandth of a millimeter (about 0.000008 inches), is colored; the cellulose in the inner side is not.

Now that the so-called two-level superficiality of the Shroud image has been described, there is a very interesting and surprising further aspect to be taken into consideration: there are, it seems, actually two imprinted images on the cloth! In fact, from the analysis of the pictures of the dorsal side of the Shroud taken in 2002, it ensued that in correspondence with the face and the hands, there is an image also in the back.

Since the Shroud image is superficial, the double image, front and back, implies a double superficiality that is, at least in correspondence with the face, an image on the cloth surface (the main one and most known) and another image, superficial, too, on the opposite side of the Sheet. Between the two sides, there is nothing.

Making an analogy, you can imagine a book with the face of the Man of the Shroud on the cover; on the back, another, even fainter, image of the same face; and in the middle, only blanks pages, without any sign of the image.

The peculiarities of the image imprinted on the sheet are not finished, indeed. The coloration of a linen thread in the image area is not continuously distributed: this means that only some fibers constituting the linen thread are yellowed, whereas others did not go through any chemical modification process.

Furthermore, the image fibers are equally colored in the circumferential direction, but they can have some color variation longitudinally.

From the optical point of view, the body image has three-dimensional features (Fig. 1.13), in the sense that a mathematical relationship can be defined between the body–cloth distance and the chiaroscuro levels of the scanned image. It looks like a photographic negative and it is not fluorescent.

From a more general point of view, the following points can be highlighted:

- The body image corresponds to a body enveloped in a sheet, in which there are noncontact zones, such as those between the nose and cheeks.
- The blood stain formation process has different characteristics compared to that of the body image; first blood stains and then the body image were imprinted. Further, there is evidence of two different kinds of corpse enveloping: The first, in terms of time, is the natural one, with the sheet that

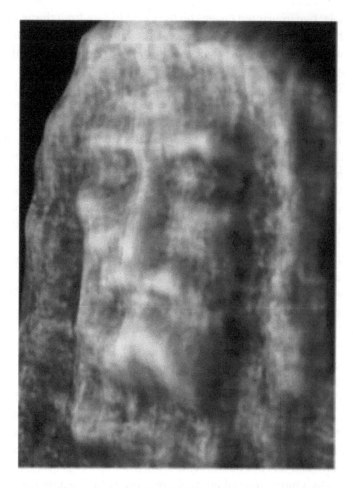

Figure 1.13 Three-dimensional elaboration of the face of the Man of the Shroud on the basis of variations of the chiaroscuro levels (Mario Azevedo).

lays down for gravity on the body, and the blood stains match with that; the second is related to the image formation and corresponds with some more flattened areas of the sheet, especially on the face area.

- The image does not show any sign of decomposition. That implies that the Man was enveloped in the sheet for a short time—according to specialists for no longer than about

40 hours. This hypothesis is supported also by the signs of rigor mortis, which usually disappear after this lapse of time.

- Finally, anthropometric studies [56] showed that, a part from the arms that had been moved during the transportation to the sepulcher, the position of the Man of the Shroud was consistent with that of a man dead by crucifixion.

1.5 How Was the Image Formed?

From the above-mentioned list about the peculiar characteristics of the Shroud body image, it is extremely hard, if not impossible, to reproduce an identical copy of the Shroud that could satisfy all those features at one time. However, several researchers proposed some hypotheses of the body image formation, and we are going to consider them in the following pages. Unfortunately, still, there are some academics that, not taking into consideration the elements previously described and all the specialized publications on the topic, are convinced that their hypotheses can be acceptable. This is the reason why the description of these hypotheses will be followed by a brief comment about their reliability.

Hypotheses considered are:

(1) artistic production;
(2) diffusion;
(3) direct contact; and
(4) radiation mechanism.

I: Artistic Production

Several researchers sustained the intervention of an artist who could have used:

- painting techniques. In that context, the attribution of the making of the Shroud has given to Leonardo da Vinci (1452–1519). This hypothesis is totally without basis since it has been proved that the image on the Shroud has not painted. Also, very reliable historical documents (analyzed in Chapters 2 and 3) testify the existence of the Shroud far earlier than the century in which da Vinvi lived.

- bas relief techniques, as, for example, that of Vittorio Pesce Delfino[19] that forged a copy of the Shroud, placing a sheet on a bas relief of metal heated to 220°C (428°F) in order to generate a scorch [109].
- a more elaborate technique using acids, as claimed by L. Garlaschelli[20] [69].

To understand the huge difficulties that a hypothetical artist would have run into in order to obtain results similar to the Shroud, especially at the microscopic level, let us see the model in Fig. 1.12. If ideally we extract a thread of the body image of the Shroud, whose diameter is 0.25 mm (0.0098 in.), and if we magnify it by about 300 times, we can think it as analogous to a bundle of drinking straws. Each straw is, in this case, a linen fiber with a diameter of 0.015 mm (0.00059 in.). From one side of the bundle we can see a dozen colored straws side by side uncolored straws. If we remove the colored film of the linen fiber, we can observe that the cellulose in the inner side is uncolored.

Now, let us think of a hypothetical artist who tries to reproduce these characteristics on a linen cloth using a simple painting technique: difficulties seem insuperable. First of all, the artist should dip the brush, not in the color, because there are not pigments on the threads, but in an acid capable of shading the linen chemically.

However, the artist has to see what he or she is painting, so the acid (usually transparent) should be pre-emptively colored, though, at work completed, he or she should eliminate any evidence of pigment, because on the Shroud there is no colorant.

Since colored fibers are side by side uncolored ones, the brush must have only one bristle with a diameter not superior to 0.01 mm (0.00039 in.). Inexplicably, the artist also has to be able to color the part of the straw in the inner side of the bundle without coloring the adjacent straws, since the color is uniformly distributed around the circumference.

Then, the acid has to be placed on the fiber just for a split second because it must have no time to act in depth and undermine the

---

[19] Professor of Antropology, University of Bari.
[20] Professor of Chemics, University of Pavia.

uncolored cellulose. Besides, the acid has to be spread uniformly along all the straw, that is, all the single fiber.

Finally the artist would have to paint in the same way all the million straw-fibers that constitute the Shroud using a microscope (not existing in the Middle Ages) and in the same time observing the Shroud from at least a 1 m distance, corresponding to a 300 m (984 ft.) distance in the straws model, because, as already explained, when we get closer to the Shroud, the body image is not visible anymore.

If these difficulties seem unsolvable nowadays, for a hypothetical, though brilliant, medieval artist they are far more problematic!

II:Diffusion

Starting from P. Vignon[21] [145], some academics proposed the diffusion mechanism in order to explain the formation of the body image. Specifically, gases developed between the corpse and the sheet could have reacted with the cloth, triggering the typical chemical reaction of the linen. R. Rogers [126] proposed a mechanism based on the Maillard reaction,[22] but all these researchers did not take in consideration that there were no signs of decomposition on the Man of the Shroud.

III: Direct Contact

Other scholars, among whom was J. Volckringer [148], proposed the body–cloth direct contact mechanism as an explanation of the image formation. This is based on the verification that after leaves are kept in a herbarium for several years, they leave a footprint on the paper they are in touch with. What is more, this footprint is negative.

Still, they do not consider that the Shroud image is evident, even in the areas where body–cloth contact was not existent, like, for example, the space between the nose and cheeks. Starting from this statement, the image formation hypotheses have to be based on a distance formation mechanism, a radiation mechanism broadly speaking.

---

[21] Paul Vignon was professor at the Institut Catholique in Paris.
[22] Maillard reaction occurs, for example, by heating sugar until brownish caramel is obtained.

IV: Radiation Mechanism

Several scholars such as F. Lattarulo [91] and G. De Liso [46] proposed a natural radiation mechanism derived from strong electromagnetic fields. A field refers to a physical entity whose characteristics vary from one point to another in space, for example, according to the gravitational field a man has a certain weight on Earth's surface that progressively decreases as the distance from Earth increases. In the case of electromagnetic fields we make reference to the different distributions in the space of electric charges produced by electrons.

They think that the body image could have been caused by an electrostatic field, perhaps related to an earthquake that could have generated a peculiar phenomenon named corona discharge[23] responsible of the creation of the image.

To explain all the particular features of the Shroud image, other researchers like O. Scheuermann [128], G. B. Judica Cordiglia [83], J. B. Rinaudo[24] [123], and E. Lindner [95] supposed the presence of a radiation source coming from the internals of the body wrapped in the Shroud, but with no success in motivating the cause physically.

Different kinds of sources of radiations have been proposed: for example, Rinaudo experimented on the effects of nuclear radiation (referring to the Resurrection described in the Gospels, but in

---

[23] Corona discharge is an electric discharge brought on by a high potential (value associated to a determined field) into a neutral fluid, usually air. It takes place when the potential is large enough to trigger the ionization of the insulating fluid but not sufficient to generate an electric arch (lightning). If ionized, the fluid becomes plasma and passes charge completing the circuit once the electric charge carried by the ions slowly reaches the ground. Corona discharge produces light, UV rays, local heating, ozone, and noise.

[24] According to Jean-Baptiste Rinaudo the oxidation and dehydration of the superficial fibers in the image areas, the three-dimensionality, and the vertical projection can be explained through an irradiation of protons emitted by the body under the effect of an unknown energy. The irradiation of protons, of 1 MeV energy, can pass through several centimeters in the air and some hundredths of a millimeter in the linen fibers. Protons, once slowed down, can produce a dehydrating reaction like that on the imprint of the Shroud. Rinaudo believes that atoms of deuterium, present in the organic matter, are involved in this phenomenon. Deuterium is the element that needs less energy to extract a proton from its nucleus, formed by a proton and a neutron. The produced protons would have formed the image, while the neutrons would have irradiated the material with a following increase of radiocarbon, which would have distorted the carbon 14 dating (see Chapter 4).

this case we are far beyond the realm of science). According to Lindner, the image has formed after electron radiation coming from the surface of the body enveloped in the Shroud. He stated that the electrons and neutrons released during the "singularity" of the Resurrection must have interacted with Jesus' Shroud, and this happened according to nuclear physics and chemical laws.

The American scholar A. Adler [4], backing the image formation as a cause of the corona discharge, supposed the presence of a ball lightning in the sepulcher, referring to a one-of-a-kind phenomenon.

The ENEA Frascati Center carried out coloration tests on the Linen on the basis of excimer lasers (emitting a UV spectrum) that gave satisfactory results [47]. However, even in this case it was not clear which kind of physical phenomenon could have triggered the laser radiation.

According to the authors, who carried out an in-depth examination of the issue both from a theoretical and an experimental point of view [52], corona discharge is the best hypothesis to explain several peculiar features of the Shroud image.

At the Department of Industrial Engineering of University of Padua (Padua, Italy), a group of scientists led by Professor Giancarlo Pesavento carried out some tests generating corona discharge on a 1:2 scale manikin covered with conductive paint and enfolded with a Shroud-like cloth. Figure 1.14 shows a Shroud-like image obtained by one of these experiments. In this case almost all the chemical-physics characteristics match to those of the Shroud, but one question remains: what could have developed a 300,000 V discharge in the sepulcher?

Summing up, the radiation hypothesis, and among these, that stating corona discharge was triggered by an intense electric field, is the most reliable because, also on the basis of experimental verifications, it allows one to obtain a result that gets close to the peculiar features of the Shroud. However, this hypothesis is not completely reliable from a strictly scientific point of view, since there does not exist yet a plausible phenomenon that should have occurred in the sepulcher that could justify such an intense but very short-lived explosion that would have created the image.

For the time being, at the scientific level, there is no solution regarding the image formation, and no indisputable explanation

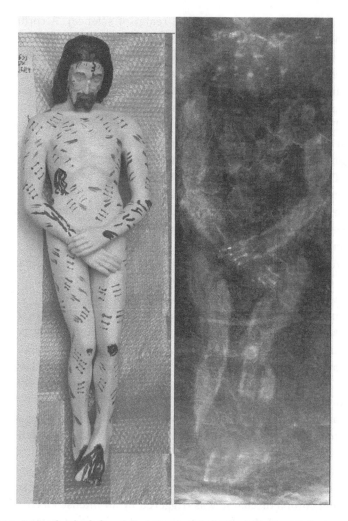

Figure 1.14 On the left, a 1:2 scale manikin built on the basis of numeric-experimental studies on the Shroud. On the right, the negative frontal body image obtained from a Shroud-like sheet enveloping a similar manikin covered with conductive paint and putting through a 300,000 V electric field. The negative pole has been, in this case, directly connected to the manikin foot. The ground consisted of a plate over the statue. Corona discharge happened in correspondence with the linen sheet.

can be given, neither on the basis of physical phenomena presently known nor with the theoretical knowledge: the issue is nowadays still unsolved and science stops here. If, to explain the body image of the Shroud, we refer to an extremely particular phenomenon, such as a miracle, we would go beyond the realm of science.

Believers or not, the Shroud brings along important causes for reflection. If the third millennium man cannot build something similar to the Shroud, since it does exist, it is necessary to ask one question: where does the Shroud come from, or better, who or what could realize a similar object? The fact that man is not able to reproduce a physical object that everybody can see and touch contrasts with the several scientific and technological discoveries, especially those developed in the last 100 years. In fact, those breakthroughs led man to deceive boastfully thinking himself to be the Lord of the Universe and to know everything. From this point of view, the Shroud is also a troublesome object because it reminds men that their knowledge is not limitless, indeed.

1.6 Blood Marks

It is stated that the image is similar to a photographic negative; on the contrary, the blood stains are obviously photopositive (Fig. 1.15) as they formed through direct contact with the corpse. UV fluorescence photographs show a pale aura around these stains.

The human blood marks match with the correct position of the wounds on the body, considering also the drapery of the cloth enveloping the corpse.

Many of these blood and serum stains are difficult to be reproduced artificially, because this blood first coagulated on a wound skin and then dissolved again by fibrinolysis[25] after the contact with the moist cloth [97].

The following description of the blood marks proves very useful for the identification of the Man of the Shroud. According to the hypothesis suggesting that this man is Jesus, a number of blood

---

[25] Fibrinolysis is a process that counterbalances blood coagulation and is in dynamic balance with it. Fibrinolysis breaks down fibrin and contrasts the platelet plug formation.

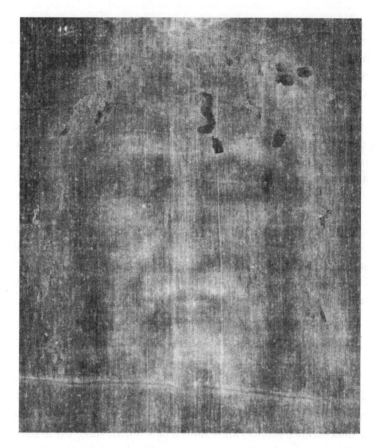

Figure 1.15 Face of the Man of the Shroud: the image is photonegative; the red blood marks are photopositive.

marks suit the description in the Gospels of the sufferings He underwent in His Passion.[26]

---

[26] The Four Gospels describe the Passion, the Death, and the Resurrection of Jesus Christ. He, betrayed by his apostle, Jude, was arrested by Romans and condemned to flagellation by the prefect of the province of Judaea, Pontius Pilate, on the Good Friday probably in A.D. 30 or 33. Immediately after, the crowd asked and obtained from Pilate the condemnation to crucifixion. He, then, was forced to carry on his shoulder the heavy cross on the way to Calvary, immediately outside Jerusalem's walls. There, he was crucified through nail embedding in correspondence with hands and feet, and he died shortly afterward. A Roman centurion, to verify Jesus' death, pierced his chest with a lance. "Blood and water" came out from the chest,

On the Shroud there are clots of many lacerated and contused wounds[27] distributed all along the body. It is probable that they have been caused by the *flagrum*, the Roman whip (Fig. 1.16 and Fig. 1.17), confirming the hypothesis of the scourging (...*after he had Jesus scourged, he handed him over to be crucified*, Matthew 27,26).

As far as the face is concerned, the following blood marks can be distinguished. Corresponding to the forehead, the scalp, and the back of the head (Fig. 1.18) there are several point- and round-shaped prints that can be correlated to the tip wounds left by a crown of thorns, confirming the hypothesis of a "coronation" (*And the soldiers wove a crown out of thorns and placed it on his head*, John 19,2). From these wounds, a number of blood flows trails down.

On both sides of the forehead blood formed two rivulets that flowed downward. They coagulated when the Man was in a straight position and not extended.

Between these flows, prominent is one in the shape like the Greek letter epsilon (ε) or a reversed "3," coming from the frontal vein (Fig. 1.15); according to the opinion of some physicians it is consistent with the blood crossing the forehead wrinkles caused by frowning under intense stress [20, 124] or, according to others, caused by a blunt object [113].

On the face, among the visible injuries, other wounds can be noticed, such as those on the lips and on the right eyelid, the latter probably caused by a scourge hit. Other blood marks on the area of the hair could correspond to the cheek zone, if we consider that the Shroud enveloped the face during the blood mark formation, whereas it took a flatter configuration during the image formation [92].

Other than blood marks on the face, the other traces to be considered are the nail puncture wounds, the chest wound caused by the lance, and the so-called *blood belt*.

---

that is, blood and serum, which appears transparent like water. He was deposed in a sepulcher near Calvary in the evening of Good Friday, and there he remained until the following Sunday, when he rose from the dead, reappearing alive to many people.

[27] In the book of G. Ricci [122] the author states to have counted more than 120 scourge marks coming from two different positions.

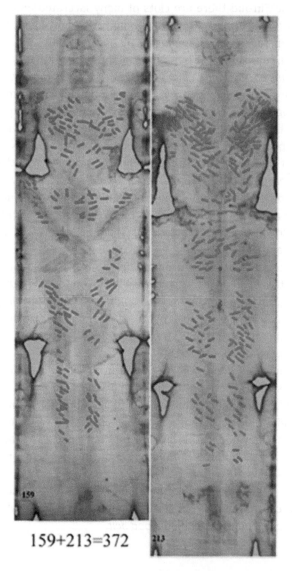

Figure 1.16 Scourging wounds have been highlighted with a dark mark; according to the first author there are 159 wounds of a scourge on the frontal image and 213 on the dorsal image, 372 wounds in all. This number would increase if some uncertain marks and all the lateral wounds, not on the Shroud, were taken into consideration.

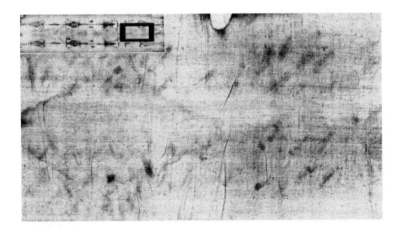

Figure 1.17 Lacerated and contused wounds caused by the Roman *flagrum*.

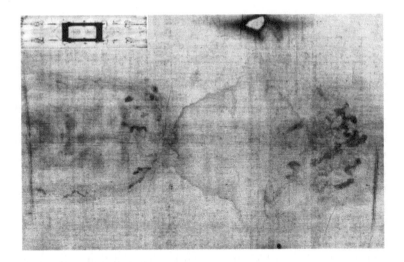

Figure 1.18 Wounds on the face and on the scalp.

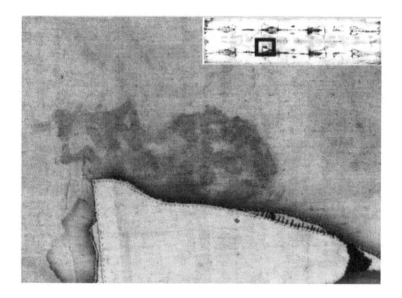

Figure 1.19   Detail of the chest wound.

Serious injuries associated to nail punctures are on the left wrist and on the feet. On the left forearm there is evidence of slight blood flow trailing down the elbow; on the right forearm a similar blood flow can be distinguished, probably related to a wrist injury (covered by the left hand) like that on the left forearm. Nail marks confirm the hypothesis that crucifixion happened (*Unless I see the mark of the nails in his hands and put my finger into the nailmarks and put my hand into his side, I will not believe,* John 20,25).

On the right side of the chest,[28] a large oval wound can be seen (Fig. 1.19), whose major axis measures 4.5 cm (1.77 in.) and minor 1.5 cm (0.59 in.). It can be deduced that it has been produced with a lance tip, which would confirm the hypothesis that the Roman centurion verified in that way that Jesus was actually dead by crucifixion (*...one soldier thrust his lance into his side, and immediately blood and water flowed out,* John 19,34). Its longitudinal

---

[28] Remember that on the Shroud imprint, it is the left side, since the image is specular.

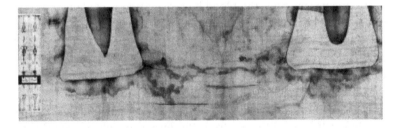

Figure 1.20 Blood belt whose origin is not clear yet: it could come from the blood flow of the chest, but more likely it derives from the blood flow of the arms or from the wounds in the kidney area.

axis is parallel to the anterior costal arches, in correspondence to the fifth right intercostal space [20].

As typical of corpses, the chest wound margins remained enlarged and well outlined: the injury must have happened postmortem. A consequence of this wound has been a severe and denser-than-normal blood loss, followed by a serum runoff. These blood marks form a 15 cm (5.91 in.) × 6 cm (2.36 in.) wide stain on the cloth. Here there is evidence of blood clots separated from serum. All this is typical of a man dead after severe blood accumulation in the chest cavity.

Another wound to be taken into consideration is the so-called blood belt (Fig. 1.20), referring to the blood flow on the inferior dorsal side of the Man of the Shroud. A number of scholars [21, 146] put in relation this imprint with the chest blood runoff. Blood would have been drained out on the Man's back because of gravity. Someone even supposed that this mark, mainly horizontal, would have corresponded to the belt that would have sustained a loincloth.

This issue has been called into question, and after various discussions with some experts of the Shroud Science Group, another kind of origin has been hypothesized. Some researchers [157] think that some blood flow occurred in a second moment, perhaps when the corpse had been laying down in the sepulcher. In moving the body to put it down on the sepulcher, the position of some limbs could have changed, causing the runoff of some blood that was not coagulated yet.

In this perspective, the blood belt formation is explained as due to the blood flow occurring after body moving. Observing the marks more in detail, it seems that they would have formed as a result of blood flow coming from two different areas, symmetrical to the body axis and located approximately in the areas of the holes caused by the Chambéry fire.

Experiments, carried out also by the first author, highlighted that this blood flow could come from the forearms, which would have dripped along the elbows.

According to another explanation, this blood flow could be caused by a serious injury in the kidney area. On the Shroud, because of its peculiar wrapping, these kinds of injuries cannot be seen, but the Man could have been seriously injured in the kidneys also during the heavy scourging he underwent.

1.7 Marks of Other Tortures

As shown, the Man of the Shroud was severely tortured, beaten with a stick, scourged, crowned with thorns, and crucified [17, 21, 146]. He shows multiple traumas: there is swelling on the forehead, on the eyebrow arch, on the cheekbones, on the cheeks, and on the nose; this also presents a deviation, possibly reflecting a fracture of the nasal cartilage. Notwithstanding the tortures, the face shows a calm serenity (Fig. 1.21).

Professor Matteo Bevilacqua, chief of Pulmonary Division at Padua hospital, wrote [58]:

> The dislocation of the right shoulder indicates a trauma suffered during the cross transportation toward Calvary, in accordance with other known evidence such as abrasions and contusions on the knees, septum fracture, and soil presence on the soles, on the left knee, and on the nose tip. It has been a violent blunt trauma on the right side of the body from behind, involving the root of the neck and the back and causing lesions of the right brachial plexus. This explains the lowering of the shoulder, the sinking of the eyeball into the orbital cavity, and limb posture with the hand and fingers flattened. The position of the right hand, whose fingers brush against the external side of the contralateral thigh, is due

Figure 1.21 Face of the Man of the Shroud in the photograph negative (G. Enrie).

to a dislocation of the right humerus, whose head is placed about 3 cm (1.18 in.) under the glenoid articulation.[29]

---

[29]The glenoid cavity is a part of the shoulder and articulates with the head of humerus, which is held in place within the glenoid cavity by means of the long head

On the left knee (and perhaps also on the right) there are abrasions due to the severe falling-down. In that position some soil [89] has also been found, which would confirm the contact with the ground. These marks are consistent with the fact that Jesus carried on his shoulders the heavy Cross on the way to Calvary.

The marks left on the face by hair and beard make even more difficult the explanation of the image formation, since they are not easy to reproduce with standard experimental techniques. This is because their apparent "softness" imprinted on the Linen, as opposed to a sweaty or oily hair image, could be reproduced if some particular conditions are applied, related to the action of a high-voltage electrostatic field that leads to hair rising [50, 59].

His beard seems partially pulled out on the right side, and long shoulder-length hair appears longer on the left side; also, the shock of hair is more marked on the left side: it is similar to "payot," the side curls of Orthodox Jews. On the right, instead, also the hair could have been pulled out. These marks confirm the hypothesis of the several blows on Jesus' face (*They spat upon him and took the reed and kept striking him on the head*, Matthew 27,30).

1.8 The Deposition of the Man of the Shroud: A Hypothesis

The mysteries related to the Shroud are still many. One of these concerns the configuration of his enveloping in the sepulcher. As shown, the scientific analysis of the body image leads to the conclusion that the Shroud enveloped a tortured man in rigor mortis, congruently with the position he assumed after his death by crucifixion, but it is not possible to add many other details.

Something more can be done if just for a moment we leave the scientific analysis aside and accept to take into consideration the parallel information deriving from the Gospels. It is known, in fact, that there is a full correspondence between what is represented on the Shroud of Turin and what is described in those Books. On the

---

of the bicep tendon. The cavity is not so shallow, and this characteristic makes the shoulder joint prone to luxation but allows 120 degrees of flexion.

basis of these premises, combining the information contained in the Holy Books to the scientific data and other materials, it is possible to try to develop, as Fanti G. did, a hypothesis about the Shroud enveloping in the sepulcher. A description follows.

From the Gospels we can deduce that, given the probable urgency of performing the burial operation at the end of Good Friday, the body was not completely buried; it was left, instead, on a stone bench for burial preparation, waiting for a complete sepulture, in order to carry out it later, the first day after Saturday. On the burial bench some dry spices were placed making the supporting surface curved;[30] then, the corpse was laid out on half of the Shroud with the face upward (Fig. 1.22); laterally bandages soaked in substances were placed to protect the body from decomposition [140]. After covering the corpse with oils and spices, the second half of the Shroud would have been just placed on the frontal side of the body and perhaps lateral bandages were used to contain the whole works. Finally, on the face covered by the Shroud Linen soaked in substances was laid to prevent decomposition. This kind of disposal would explain the reason why no lateral images are visible: the rolled bandages would have prevented their formation.

At this stage things become more complicated, because on the Shroud it is possible to observe some thin blood marks in correspondence with the hair. These marks are not consistent with the blood clots in the hair; they are, instead, consistent with the wounds on the skin. If the Shroud had enveloped the face, these blood marks, because of the warping caused by the enveloping, would have corresponded to a position near the ear. Why, then, the stains on the hair? It can be deduced that these blood marks have been printed on the Shroud in a different moment than that of the body image formation, blood before, the image after. Thus, it is probable that the enveloping configuration was different in the two moments.

The blood marks would have been formed when the Shroud partially wrapped the body and the face, and by consequence, some of these, evident on the hair, would be due to a cylindrical distortion

---

[30]Otherwise trough, hollow.

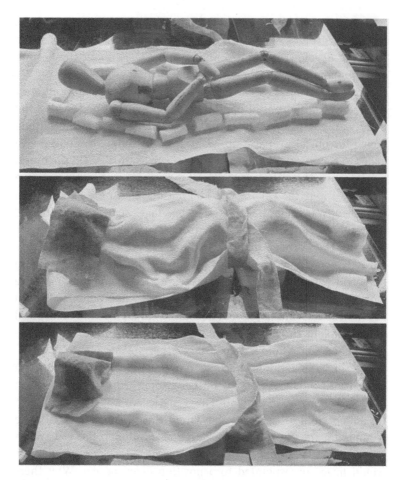

Figure 1.22 The different stages of the Man of the Shroud enveloping. (1) The Man of the Shroud is laid down on a stone bench; the corpse is laid out on half of the Shroud; laterally, rolled bandages soaked in aromatic substances are placed. (2) The second half of the Shroud is placed on the body, and the face is covered with a cloth soaked in aromatic substances. Probably the Shroud is partially closed with bandages in correspondence with the legs. (3) After the energy explosion (the not yet explainable scientific phenomenon) the superior half of the Shroud, depleted of the corpse, falls down for gravity, leaving the *sudarium*, stiffened because of aromas, in the same position in which it was when it enveloped the face.

caused by the partial wrapping of the sheet, which also touched the area of the cheeks near the ears.

This wrapping would have lasted about 40 hours, the time needed for the development of fibrinolysis[31] or dissolution of the blood clots, as some researchers supposed [28].

There is no evidence of decomposition, especially in the area correspondent to the orifices, [20, 146, 157] otherwise the body image would be stained, mainly in correspondence with mouth and nose (Fig. 1.23); by consequence, the Man of the Shroud could not have been enveloped in the cloth for more than 40 hours.

After this time, a body image would have been formed through a sudden energy explosion that also would be responsible for the repositioning of the sheet [50, 59]: in the corona discharge hypothesis, the moustache, beard, and hair would have stood up in all directions because of the electric discharges,[32] this would have caused a flattening of the Shroud that moved from the cheek area, adjacent to the ears.

Hair, hoisted owing to electric discharges, would have occupied this area, imprinting its image in correspondence with the above-mentioned blood stains. Moreover, the hair on the Shroud appears soft, as typical of electrification and not characteristic of the packing effect of body fluids and aromas.

In agreement with a Canadian scientist [116], it is necessary to consider also the fact that the hypothetic burst of energy would have heated the air between the human body and the cloth, causing a pressure increase that would have repositioned the Shroud, flattening it, during the body image formation. Also the American researcher J. Jackson [81] theorized a more flattened configuration of the cloth, hypothesizing a man who had become mechanically transparent.[33]

This hypothesis has been expressed after having observed both the lack of flaws on blood marks and the lack of clot break that could

---

[31] See footnote 25 on p. 31.

[32] Discharges would be generated by the hypothetical electric field that, through the so-called corona discharge (see footnote 23 on p. 28), would have formed the body image.

[33] Though this hypothesis has been formulated by world famous physicists, it is out of the realm of traditional science.

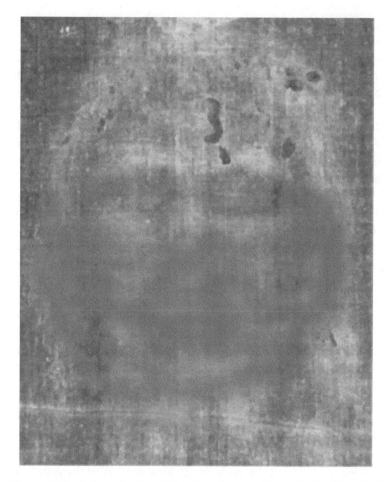

Figure 1.23 A hypothesis of the face image of the Turin Shroud if decomposition gases would have interacted with linen.

be a consequence of a hypothetical mechanical transportation of the corpse out of the Shroud. According to this assumption, the sheet collapsed into the body region during the time of image formation.[34]

The cloth partially enfolding the face, being stiffed because of ointments, would have stayed in the original position, making the

---

[34] Resurrected corpse? This question is out of the realm of traditional science, since Resurrection is a unique phenomenon.

Shroud disposition extremely unusual. Maybe this is the reason why John the apostle, taking part in the procedures of corpse preparation on Good Friday, after entering the sepulcher the following Sunday morning "... saw and believed" (John, 20,8).

This is what we can retrace on the basis of the present historical and scientific knowledge related to the Man of the Shroud, to the sheet enveloping, and to the particular environmental conditions that could have reproduced the extraordinary body image that to this time is still unexplainable.

The identikit has been traced, but it is obviously not complete, since no detective is allowed to go back out to the "crime scene" to analyze it with modern techniques. The following pages will try to produce further details to the Shroud enigma, also referring to the dating issue, in order to allow the reader to better understand this captivating object.

Chapter 2

Historical Evidence

The history of the Shroud is very convoluted and not easy to retrace in all its steps. In fact, whereas it is quite simple to find documents and evidence of an object exposed, contended, and conserved in the past centuries near to the present, it is indeed more difficult to trace it back from the remote past, when evidence fades away and also mixes up with legends. The 2000-year history of this relic includes not-easy-to-investigate dark periods.

This is the reason why this chapter begins with the documented history of the Shroud in more recent times, whereas various hints related to its first centuries follow. The chapter ends with a historical description of the outstanding scientific developments that occurred from the past century to present days. A very interesting historical source is, then, numismatic analysis. Since it is a very wide subject, it will be developed in the next chapter.

2.1 Recorded History

The first, almost universally recognized, historical documents related to the Shroud date back to a big dispute that happened in the second half of the fourteenth century between Geoffroy de Charney,

The Shroud of Turin: First Century after Christ!
Giulio Fanti and Pierandrea Malfi
Copyright © 2015 Pan Stanford Publishing Pte. Ltd.
ISBN 978-981-4669-12-2 (Hardcover), 978-981-4669-13-9 (eBook)
www.panstanford.com

Figure 2.1 Pilgrim medallion showing the Shroud (Museum Cluny, Paris).

the then-owner of the cloth; the canons of Lirey, who received it in order to conserve it; Pierre d'Arcis, Bishop of Troyes; King Charles VI; and Antipope Clement VII.

When de Charney was still alive, the Shroud was publicly exposed in Lirey; this is also documented by a lead medallion (Fig. 2.1), found in the Seine river in Paris in 1855, portraying the Shroud and showing the coats of arms of the families of de Charney and de Vergy. Also, a document dated 1389 and written by d'Arcis states that the Shroud had been publicly exposed about 34 years before, around 1355.

It is this text, a letter to be more precise, that is used as one of the most important evidence against the Shroud's authenticity. The Bishop of Troyes, Pierre d'Arcis, irritated for the pompous ceremonies and because of the pilgrims, who being attracted by the Shroud, defected Troyes, wrote a long letter to Antipope Clement VII, in which he claimed that experts theologians and trustworthy men ensured him that the Shroud of Lirey could not be authentic because, if an imprint had been visible on Jesus' burial cloth, the

Gospels would have certainly mentioned it.[1] Furthermore he also claimed that a painter confessed to forging it. However, the bishop did not have documents or evidence corroborating his claims. By the other side, de Charney complained to the antipope about the bishop's behavior.

Clement VII, fed up with that controversy, on January 6, 1390, issued a Bull and two additional letters discussing the Shroud. The three points debated were:

(1) "Public showing of the relic is authorized, but ceremonies censored by the bishop are forbidden. Furthermore, during displaying, it has to be announced loud and clear that it is not the true Sudarium of the Lord, but a *pictura seu tabula*,[2] a picture copy of the Shroud (A/N: the expression clearly indicates a work made with hands).

(2) d'Arcis was forbidden to oppose the exposition of the cloth, so long as it was carried out in the manner prescribed.

(3) The ecclesiastical officials of Langres, Autun, and Châlons-sur-Marne are invited to give forth all the prescriptions related to the Shroud and to make sure that they will be observed."

It is very meaningful that the expression *pictura seu tabula* written in the January 6, 1390, Bull had been substituted with the simple indication of *figura seu rapresentacio*[3] (that results in all the other papal documents) attached to the copy of Regesto Vaticano.[4]

---

[1] Actually, it is likely that the Gospels did not mention the body imprints on the Shroud because they were not visible by that time. The hypothesis sustained also by the authors, based on the image formation process caused by an intense electric field (called corona discharge), calculates that the image consequently formed would have taken shape slowly in the course of time (months or years). By consequence, it can be easily thought that the Evangelists could only have observed the blood marks imprinted on the Shroud.

[2] "Painting or painted table." The term *tabula*, or painted table, indicates the high ignorance degree of the person writing this text on the relic, since, as it is well known, it is a linen cloth and not a table.

[3] Figure or representation.

[4] The Regesto Vaticano (T/N: Vatican Register) for the Calabria is a 12-volume work edited by Francesco Russo in the past century containing the files of Early Middle Age documents and Byzantine codes (about 700 registers and 1000 Bulls) referring to Calabria (T/N: Calabria is a region in Southern Italy forming the "toe" of the Italian peninsula).

The correction is dated May 30, 1390, the eve of the last Bull of June 1, 1390. Through this Bull, indulgences were granted to visitors of St. Mary's Church in Lirey, in which *venerabiliter figura seu representacio sudarii Domini nostri Jesu Christi*[5] was conserved.

However, even this considerable historical documentation, certifying that the Shroud was in France in the second half of the fourteenth century, has been called into question by the supporters of the idea that the Shroud had been made by Leonardo da Vinci. They simply claim that the cloth exposed in Lirey was just one of the many copies of the Shroud disseminated in France at that time! Taking it to extremes, we could declare that the Shroud now conserved in Turin is a copy of another relic, perhaps destroyed in the past. Nevertheless, we do not need to be Sherlock Holmes to realize that this thesis cannot be realistic, since da Vinci lived from 1452 to 1519, which is more than one century after these historical happenings, but people's fantasy has no limits, especially for those who want to deny at all costs the historical recorded evidence. There is a reason to that, actually: the Shroud is indeed a troublesome relic for many people, and it is easy to find, among them, scholars who attempt to rejuvenate its history as much as possible in order to try to demonstrate that this sheet is too recent to be considered the cloth that had enveloped Jesus Christ's body.

If the Shroud had been an archeological find attributable to a great emperor of the ancient history, the debate would not have been so high pitched and very less evidence would have been necessary to reach a universally accepted identification. Instead, this is a burning issue, because if there had been a certainty that the Shroud enveloped our Lord, many people could have been forced to change their mind about several aspects, especially those related to the Supernatural. Here is why, to validate their theses, some people lost their historical objectivity, not considering some recorded evidence and even real documents.

Getting back to the historical traces, until 1418 nobody heard about the Shroud anymore. In that date, Count Humbert de La Roche issued a receipt to the canons of Lirey listing all the relics he took for

---

[5] It means that "the figure or representation of the Shroud of our Lord Jesus Christ is respectfully" conserved.

safekeeping because of the disorders happening in that region. He had married Marguerite, granddaughter of de Charney, and became Lord of Lirey. He died without returning the relics, among which was *"a cloth, on which is a figure or representation of the Shroud of our Lord Jesus Christ."*

When the canons wanted to have their goods back from the widow Marguerite, they had to fight through the court. However, as far as the Shroud was concerned, she declared she was not bound to her husband's signature, claiming that the relic was of her own property, acquired from her grandfather as a prize of war. Marguerite handed the Shroud over to Anna of Lusignano-Châtillon (also known as Anna of Cyprus), wife of Louis, Duke of Savoy, on March 22, 1453. In 1502 the relic was stored by Philibert II in Saint Chapelle of Chambéry castle.

During the night of December 3–4, 1532, a fire burnt up in Saint Chapelle. It seriously damaged the Holy Linen, but the two frontal and dorsal body images remained almost untouched. Two years later, the Poor Clare nuns repaired the sheet by sewing about 30 patches and a linen cloth attached to the entire back of the Shroud for support, called "Holland cloth".

In 1506 Pope Julius II granted a specific liturgy for the holiday of the Holy Shroud, symbolically fixed [68] on May 4,[6] with its proper Mass and office. From then, the Shroud, except for short pauses imposed by events, stayed in Chambéry until 1578, when Saint Charles Borromeo, archbishop of Milan, wanted to make a pilgrimage to Chambéry to venerate the Shroud in fulfillment of a vow. To spare him the rigorous trip over the Alps, Emmanuel Filbert, Duke of Savoy, brought the Holy Linen to Turin.

On June 1, 1694, the Shroud was placed in the chapel built by the architect Guarino Guarini, adjacent to the Turin Cathedral. That year, Blessed Sebastian Valfrè[7] strengthened the patches of the

---

[6] It is noteworthy the fact that Pope Julius II approves not only the veneration of the relic, but also its adoration (perhaps because according to his opinion it contained blood, thus part of the body of Jesus Christ).

[7] Sebastian Valfré (1629–1710) from Verduno (T/N: Verduno is a country, then in the Duchy of Savoy, now in the province of Cuneo in the Piedmont region of Italy), during his stay in Turin as chaplain of the Savoy Court and master of ceremonies of Victor Amadeus, in 1694 could "… attend to the renewal of the Veils of the Holy Shroud …," and also sew it partially.

Shroud. From then on, no other restoration interventions have been documented till 2002, when all the sixteenth-century patches have been removed and the supporting Holland cloth has been replaced.

2.2 Traces from the Past

The Holy Scripture does not convey any physical description of Jesus of Nazareth and the prohibitions of the Ancient Law[8] prevented the disciples from carving the image in paintings or statues, though there are some that have been ascribed to Saint Luke or to Nicodemus by legends.

If we extend the research beyond the recognized historical documents, what are the historical traces of the Shroud? First of all, we need to choose the period of time we want to analyze, and it stands to reason that we want to go as far back in time as possible, starting to collect hints from the most known texts narrating what happened after Jesus' death on the Cross: the Canonic Gospels.

On Friday, April 3, in the year A.D. 33 or, according to someone, on Friday, April 7, in the year A.D. 30, Jesus' body was enveloped in pure white linen. In fact, the three synoptic Gospels narrate that the body of Christ was enveloped in a sheet,[9] whereas John the Evangelist writes, "They took the body of Jesus and bound it with burial cloths along with the spices, according to the Jewish burial custom" (John 19,40; 20,6).

The expression "burial cloths" is a more generic English translation of the Latin term *linteamina* (linteàmina) and of the Greek word οθόνια (othónia) in the plural in both the Latin and the Greek versions of the above-mentioned evangelic passage and that indicate linens. John the Apostle, who supervised Jesus' burial, chose the plural term to indicate all the linens used for the sepulture: the Shroud, a sudarium, and the rolled bandages soaked in aromas placed on the sides of the corpse for better conservation. The final burial would have taken place after Easter Saturday.

---

[8]Exodus 20,4; Deuteronomy 5,8: "You shall not carve idols for yourselves in the shape of anything in the sky above or on the earth below or in the waters beneath the earth."

[9]As already specified, *sindón* (σινδών) in Ancient Greek; *síndon* in Latin.

On Easter morning this sheet was found empty and it could have been picked up by Joseph of Arimathea and given to the Virgin Mary in order to keep it safe. This eventuality is described[10] in the revelations written by M. Valtorta [141]. This writing, though lacking of any scientific value, can offer a starting point for the historical reconstruction.

The apocryphal gospels talk about the burial cloths of Jesus; Jerome reports[11] a passage written in the Gospel of the Hebrews (second century) that states that Jesus gave a linen cloth to the servant of the priest preceding his appearance to James.

During the early centuries after Christ, the direct references related to the Shroud are fragmentary and often history merges with legends, for example, the face of Christ on the *Mandylion* of Edessa[12] (a depiction of the face of Jesus on a piece of cloth).

In the beginning of the history of the Church, the Jesus burial sheet was probably kept hidden for several reasons: first of all, it was a very precious "memory," having enveloped He who sacrificed Himself on the Cross. Furthermore, Christians feared that someone could seize and destroy it: the Hebrews, in compliance with Mosaic Law, considered everything that had touched a corpse as

---

[10] Being a "revelation," the following text is not verified; but on the other side it is perhaps the only available source that, moreover, is not in contrast with any other information. *And Joseph* [of Arimathea] *hands Her* [Mary] *a bulky roll, that enveloped in a dark red cloth, he had held so far concealed under his mantle. ...<< ... It is the clean Shroud in which the most pure Lord was enveloped after His torture and after the purification ...of His members soiled by His enemies, and the summary embalming. ... Joseph took both away from the Sepulchre and brought them to us at Bethany ... Then ... we gave You the first Shroud, and Nicodemus got the other ... >>.* [Lazarus goes on] *<< ... With regard to the Shrouds,... I was thinking of making a statue of Jesus crucified ... I will use one of my gigantic cedars of Lebanon—and of concealing one of the Shrouds inside it, the first one, if You, Mother, will give it back to us.... [On it] the image is so confused that it is difficult to distinguish it.... But the other one, the second Shroud, which was on Him from the evening of Preparation Day until the dawn of the Resurrection, must come to You. ... And you must be informed that the more the days passed, the more clearly His image appeared, as He was after being washed.>>* (pp. 549–461).

[11] *De Viris Illustribus* (On illustrious men) is a collection of short biographies of characters of the New Testament and of the early centuries of Christianity, written in Latin by Eusebius Sophronius Hieronymus (Saint Jerome). He completed his work in Bethlehem around 392–393.

[12] Edessa is now the city of Urfa, in South Turkey.

impure; and not Hebrews judged the punishment of crucifixion as ignominious. The reasons why the protectors of the Shroud wanted to keep it hidden are then clear.

Nino, who evangelized Georgia under the Constantine empire (306–337), inquired after the Shroud to Niafori, his master, and to other Christian scholars of Jerusalem. He learned that the burial cloths had been for some time in possession of Pilate's wife, and after, they were handed by Luke the evangelist, who stored them in a safe place known only to himself.

In the fourth century, in Edessa there was the certainty that the city owned an image of Christ, created by God and not produced by the hands of man. It is said that when the image was shown it was folded in eight layers: the result of creasing the Shroud in this way gives a long rectangle with the head in its center, without a neck. This is exactly the same image shown in the copies of the Image of Edessa.

A tenth-century codex, named *Codex Vossianus Latinus Q69*, found by Gino Zaninotto [156] and preserved in the library of the Rijksuniversiteit of Leiden (the Netherlands), sheds some light over the obscure centuries of the Shroud in Turkey: it contains an eighth-century account, coming from the Syriac area and translated by Smira archiater, saying that an imprint of Christ's whole body was left on a canvas kept in the cathedral of Saint Sophie at Edessa. Also in the Second Council of Nicaea (787), debating the veneration of holy images, there were been various talks about the Image of Edessa, not produced by the hands of man.

The great fame of the face of Jesus imprinted on a canvas spread both in the East and the West. In 944, the Byzantine army, during an offensive against the Arab Sultanate of Edessa, came into possession of the *Mandylion* and took it solemnly to its capital. Another confirmation of the arrival of the Shroud in Constantinople is given by the image painted by John Skylitzes[13] in his homonymous codex in which is depicted the arrival of the relic in Constantinople in 944 (Fig. 2.2).

---

[13] Biblioteca Nacional de Madrid (codex 26,2), the page came from a copy of the Chronicles of John Skylitzes, a Byzantine historian who lived under the reign of Alexius I Comnenius (1081–1118).

Figure 2.2   The arrival of the Shroud to Constantinople in A.D. 944 (Madrid Skylitzes, Biblioteca Nacional de Madrid).

In Constantinople, the Shroud was probably opened and folded in four layers, showing not only the face but also part of the chest. Hence the birth, during the twelfth century, of *Imago Pietatis*[14] (Fig. 2.3), the representation of dead Christ at half-length protruding from the sepulcher bolt upright with crossed hands [40, 111].

Among the historical information of the Shroud in Constantinople it results that Louis VII, King of France, during his visit, venerated it in 1147 and that in 1171 Manuel I Comnenus showed to Amalric, King of Jerusalem, the relics of the Passion, among them, the Shroud.

A speech ascribed to Constantine VII Porphyrogenitus [96], Byzantine Emperor from 912 to 959 and expert at painting, gives interesting information about the *Mandylion* of Edessa, suggesting a new explanation of the image due to "a liquid secretion" without colors or paint art.[15]

---

[14] Image of Pity, the representation of pity.

[15] *How the form of a face could be imparted onto the linen cloth from a moist secretion with no paint or artistic craft, and how something made from such a perishable material was not destroyed with time, and whatever else the supposed investigator*

Figure 2.3 Russian icon of Imago Pietatis covered with metal riza, the nineteenth century.

In support of the identification of the Image of Edessa with the Shroud, Gino Zaninotto [156] found in the Vatican Archive *Codex Vatican Graecus 511* (tenth century).[16] (on a cloth) *of Christ, which*

_____

of natural causes is wont to enquire into with curiosity: these questions he should yield to God's inscrutable wisdom.

[16]The complete title is *Sermon by Gregory the Archdeacon and Referendarius of the Great Church at Constantinople*, about how incredible things are not subject to the laws of praise, and about how three patriarchs have declared that there is an image.

was brought from Edessa 919 years afterward by the zeal of a pious emperor (Romanus I Lecapenus), *in the year 6452* (A.D. 944). *Lord bless us.* Translation from Greek by Marc Guscin, "The Sermon of Gregory Referendarius," 2004. This codex lists the colors used for drawing icons faces[17] and states that the image has not been reproduced with artificial colors, as it is only "reflection."

Gregory explains the image as follows:

> This reflection, however—let everyone be inspired with the explanation—has been imprinted only by the sweat from the face of the originator of life, falling like drops of blood, and by the finger of God. For these are the beauties that have made up the true imprint of Christ, since after the drops fell, it was embellished by drops from his own side. Both are highly instructive—blood and water there, here sweat and image. Oh equality of happenings, since both have their origin in the same person.

Zaninotto underlines the unexpected element in this work, also not required for the structure of the story: the detail of the side wound is senseless since the tradition sets the image formation after the crucifixion. The Image of Edessa, actually, did not depict only a face but also part of the breast at least up to the chest. By consequence we can observe that the image described by Gregory the Referendarius is the relic and not a common depiction of Christ, otherwise the side wound detail would have no reason to be mentioned.

From the numismatic point of view, in the next chapter it will be shown in detail that during the Byzantine Empire the Shroud should have been deeply known, because it inspired the representations of almost all the faces of Jesus carved on the coins minted in that period.

In 1204, Robert de Clary, chronicler of the Fourth Crusade, writes in his work *La conquête de Constantinople* that before the fall of Constantinople by the Occidental side (April 14, 1204), a Sydoine (T/N: a shroud) was exhibited every Friday in the Church of Saint Mary of Blachernae, and on that cloth, the figure of Christ was clearly

---

[17] It shows a sufficient knowledge of painting art and a discrete ability in discovering a potential fake.

visible. *But*—he adds—*nobody knows yet what happened to the Sheet after the conquest of the city.*

So, the Shroud disappears from Constantinople, and it is likely that its cover-up was provoked by the fear of being excommunicated for relic stealing.

Some hints lead to the idea that the Shroud was brought to Europe and kept by the Templars for one and a half centuries. Other clues imply that during its transportation in Europe, the Shroud passed through Greece.

In a letter sent to Pope Innocent III in 1205 by Theodore Angelus Comnenus, ruler of Despotate of Epirus and grandson of Isaac II, Byzantine emperor when the city was ransacked by the Latin crusaders, it is stated that the Shroud was in Athens. In France, in 1307, the Templars were persecuted and arrested by Philip the Handsome, who confiscated all their wealth.

The face venerated by the Templars (Fig. 2.4) has similarities with the face of the Shroud. The original has never been found, but a copy of the image they worshipped came to light in 1945 in Templecombe, England, in one of their ancient headquarters: it is a bearded face with blurred outlines. It is painted on an oaken table,

Figure 2.4 The face venerated by the Templars, Templecombe (England), 12th–14th century (I. Wilson).

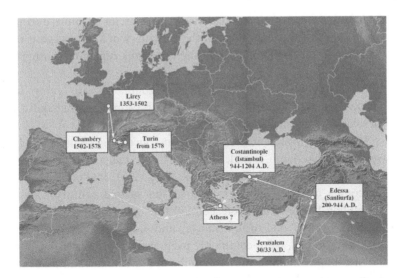

Figure 2.5   Presumable historical route of the Shroud during the centuries.

dated between the twelfth and fourteenth centuries, that could be a lid of a case. It could be deduced that the case contained the Shroud and that the Templars kept it there in the years of concealing.

To sum up, referring to Fig. 2.5, the more reliable historical route of the Shroud is the following: from year A.D. 30 or 33 the Shroud was in Jerusalem in the Holy Sepulcher; in the lapse of time between approximately A.D. 200 and 944 it was exhibited several times and also conserved for various decades into the city walls of Edessa. From 944 it was publicly displayed in Constantinople till the Fourth Crusade ransack in 1204.

Since then, it is not clear who seized it, and some historians maintain that it passed through Athens. The *Codex Pray*, a collection of medieval manuscripts, kept in the National Széchényi Library of Budapest, named after György Pray, who discovered it in 1770, testifies that the person who wrote it had to have seen the Shroud. Then, the relic appears in Lirey, France, in 1355, and in 1502 it was brought to Chambéry, in which in 1532 it was affected by the well-known fire that severely damaged it.

Since 1578 the relic was brought to Turin, and there it remained to the present day, a part from some short periods of wars and disorders in which it was carried to other places; for example, during World War II it was concealed in the Sanctuary of Montevergine, in the province of Avellino (T/N: in the Campania region of Southern Italy).

After having rebuilt the historical route, we are going to analyze some iconographic traces of the first centuries that show a correlation with the Shroud image.

During the first centuries of Shroud concealing, some symbols were used, like lamb, bread, and fish.[18] The image of "Eucharistic fish" can be observed, for example, (Fig. 2.6), in Rome, in the catacombs of Saint Callistus (end of the second century). Also, the human figure of an adolescent, of the "good shepherd," of the "miracle worker," and of the "master and judge," usually based on the classic model, were commonly used as symbols. Christ healing the bleeding woman in the catacombs of Marcellinus and Peter in Rome is one of these kinds of examples.

One of the first representations of a bearded Christ surviving till the present days is dated A.D. 290–310 and Jesus blessing is depicted with a beard and bushy hair. It is conserved in the National Roman Museum—Palazzo Massimo alle Terme (Fig. 2.7). In fact, there is a fragment of marbled plate with scenes from the New Testament. This indicates that already in the end of the third century some information about the physiognomy of Jesus on the basis of the Shroud image reached even Rome. The young and beardless Jesus become bearded and with long hair.

Other examples of a majestic and bearded Christ can be found on the sarcophagi in Lateran Museum, Saint Sebastian outside the walls (around A.D. 370), Arles (before A.D. 370), and Basilica of Sant'Ambrogio in Milan (380–390). Since then, throughout the centuries this kind of representation was not abandoned anymore and perhaps there was a good reason to keep it: the representations of Christ, even in the present days, hark back to the Shroud image.

---

[18] "Fish" in Ancient Greek is ἰχθύς (ichthús), ΙΧΘΥΣ in capital letters, and Christians used it to recognize themselves in the period of persecutions. The word "fish" results from the initials of the words Ἰησοῦς Χριστός Θεοῦ Ὑιός Σωτήρ (Iesoùs CHristòs THeoù Yiòs Sotèr)—literally Jesus Christ, God's Son, Savior, that is, Jesus Christ, the Son of God, our Savior.

Figure 2.6   Christ healing a bleeding woman, the catacombs of Marcellinus, and Peter (Rome), third century (M. Paolicchi).

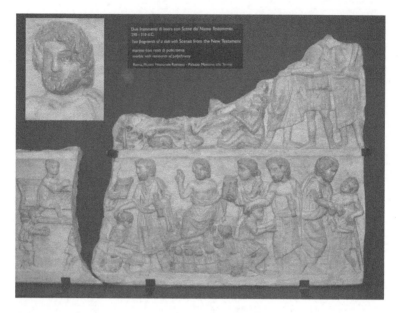

Figure 2.7 Marbled representation of a blessing Jesus with a beard and bushy hair. A.D. 290–310. National Roman Museum—Palazzo Massimo alle Terme.

P. Vignon [145] and many scholars claim that the artistic depiction of Christ has to depend on the Shroud, in the sense that there is a resemblance between the classic model of the face of Jesus with a beard and the Shroud image.

It is known that Byzantine iconographers left no space to imagination; thus, their paintings should have been in compliance with the severe rules handed down for centuries and millenniums.

In *Painter's Manual of Dyonisus of Fourna*[19] these rules are described in detail, and among them, as far as the Savior is concerned, he wrote:

> The body of God in human form is three cubits tall; his head is slightly inclined, and his gentleness is particularly apparent. He

---

[19] *Painter's Manual of Dionysius of Fourna: An English Translation* (from the Greek), Leningrad Dionysios of Fourna (Paul Hetherington, trans.), Sagittarius Press, London, 1974.

has well-formed eyebrows, in keeping with the rest; fine eyes, a fine nose, a corn-colored complexion, and slightly auburn, curly hair on his head. His beard is black, and the fingers of his spotless hands are long in proportion. His character is simple, like that of a child, to which he was similar when alive, and which is that of perfect man.

It is evident that many of the characteristics above described derive from the Shroud, particularly the long fingers proportioned to the hands, already assumed from the Shroud image in the first centuries after Christ.

At the same time of the discovery of the *Mandylion* in Edessa, a particular kind of Jesus' portrait took root in the East. It is the majestic Christ, with beard and mustache, that is depicted in many different ways, such as the *Pantocrator*[20] in the post-Byzantine age and whose features will be substantially reproduced till this day. According to the Eastern Church, the only reliable portrait of Jesus is based on the Image of Edessa, the *Mandylion* (Fig. 2.8).

The depiction of the face of Christ, starting from the sixth century, shows some asymmetrical and irregular features, hardly to be ascribed to the artist's imagination. In particular, we can observe:

- long, bipartite, and often asymmetrical hair along the face; a lock of short hair, with different ends, on the forehead;
- pronounced eyebrow arches; a V-shaped mark on the bridge of the nose;
- big and deep eyes, wide open and big orbits; long and straight nose;
- pronounced cheekbones;
- a little mouth, not hidden by a mustache, which is often droopy; a hairless area between the lower lip and the beard; and
- a not too long and asymmetrical beard.

The first example of this kind of iconography is the Pantocrator, conserved in Saint Catherine's Monastery, which was built in A.D. 500 by order of Emperor Justinian I at the foot of Mount Sinai, Egypt

---

[20]"Pantocrator" means "the Almighty," the "Lord of the world."

Figure 2.8 Mandylion icon, Russia, 18th century.

(Fig. 2.9). The Shroud bibliography found many points of congruence between the Pantocrator icon and the Shroud face, demonstrating that who had painted the icon surely saw the Shroud (Fig. 2.10).

Recently, Professor Fanti discovered that, observing the eyebrows of the Pantocrator of Saint Catherine, there is a noticeable difference between the right and the left that contributes to a diversification of the facial expression. Some scholars commented on this dissimilarity, affirming that the iconographer wanted to represent two faces of the Lord in the same icon, one more severe (the left half) and one more gentle (the half on the right).

Figure 2.9 Pantocrator in Saint Catherine's Monastery in Sinai: In the center is the original image; on its sides the symmetrical images of the right and the left side of the face, highlighting the asymmetry.

Figure 2.10 Comparison between the face of Christ Pantocrator in Sinai (to the left) and the face of the Man of the Shroud (to the right).

Actually, this asymmetry of the left eyebrow exactly corresponds to the same irregularity of the Shroud face. It has, in fact, a more raised left eyebrow and with the same sharp outline due to a swelling caused by one of the many blows he came under.

Among the first depictions of the Christ face, the one painted on the vase from Emesa, the modern Homs in Syria, dated between

Figure 2.11 Face on the vase from Emesa [82], modern Homs, Syria (sixth and seventh centuries).

the sixth and the seventh century, is a good example. On this vase a man with a mustache, beard, and Shroud-like long curly hair can be observed (Fig. 2.11). Furthermore, the cheekbones are pronounced, the right more than the left one, and the area under the inferior lip is hairless. These features are analogous to those of the Shroud face.

Also the face of Christ on the cameo shown in Fig. 2.12, sixth century, has the same above-mentioned features of the Shroud face.

Before than the iconoclast debate in the eighth century forbade all the Jesus Christ depictions, the representation of the Savior under the guise of a lamb was pretty common and used for several centuries; for example, in the Soisson Cathedral in France, built in

Figure 2.12 Picture of the face of Christ on a cameo once conserved at the National Museum in Vienna, today regained by the ancient Polish family Lanckoronoski (sixth century). The cameo was found by the Viennese sindonologist Gerturud Wally in the museum archives.

Figure 2.13 Depiction of the lamb enveloped in the Shroud. Soisson Cathedral, France, 1200 (photo Traudl Wally).

1200, an interesting image of the lamb enveloped in the Shroud borne by two angels is visible. This is also another sign that the relic of Turin was already known in those years (Fig. 2.13).

Also the orthodox Cross, depicting one of the three horizontal crossbeams, the lowest, slanted, is in compliance with the representation of Christ on the Byzantine coins (Fig. 2.14): perhaps it was thought that Jesus was lame [34].

It is likely that this belief is derived from the fact that the body image of the Shroud shows the feet of the Man as deformed. They remained fixed in that position because of rigor mortis, after the Man of the Shroud had been un-nailed from the Cross.

Figure 2.14 Left image: Byzantine cross with the third crossbeam slanted. Right image: Golden silver cross. 12 × 9 cm ascribed to Charlemagne. 12th century. From Aachen Cathedral Treasury, Germany, http://www. bildin-dex.de/dokumente/html/obj20460173#—home. Picture by Marburg, registration Nr. 1.086.066; (9 × 12).

In fact, in the Shroud image the left foot[21] is in an anomalous position, which can wrongly lead one to think a lame man if no references are made to Cross nailing and to the body enveloping into the sheet that occurred afterward.

The anomaly of the Byzantine depicting the right foot thinner and right angle rounded compared to the left foot is also very common in the most recent Eastern iconography.

Also the twelfth-century cross ascribed to Charlemagne and conserved the Aachen Cathedral Treasury, in Germany, seems to confirm this (see again Fig. 2.14). In fact, even if the third slanted crossbeam is not depicted, the right foot is inclined without any

---

[21]Apparently the right foot, since the image is specular. The knowledge about the image formation in that age was extremely insufficient, and probably not so many people thought that the Shroud image was specular, being consistent with the corpse enveloping; so the left foot seemed the right; thus it is represented in this way in many iconographic depiction, also numismatic.

logical explanation if the Byzantine tradition recalling the Shroud is not taken into consideration.

From this overview about the face of Christ iconography of the first centuries, beyond the obvious interest of different populations in representing the God-made-man, the close connection of the typical features of these depictions with those of the Shroud face clearly results, demonstrating that the relic had often been a model for the representations of Christ.

2.3 The Scientific Century

The first photograph taken by the lawyer Secondo Pia in 1898 blazed a trail for an intense development of scientific research, prolonged throughout all the twentieth century. The real innovation was that these detailed photographs allowed the scientists to carry out their studies in their own laboratories instead of being forced to observe the Shroud in person.

Exactly 100 years after, at the end of the 1998 Turin Symposium for the Study of the Shroud of Turin, the director of the Turin Center, Bruno Barberis, probably after the "scientific disaster" caused by the well-known carbon dating analysis made in 1988, during the dinner with the speakers declared that from that moment on, the scientific research on the Shroud was official closed.

Obviously, it was not possible to stop the study of all the scientists working on the samples taken from the cloth, but officially, the Turin Center stopped any new research. In fact, also the important results obtained by the first author presiding a research project carried out by the University of Padua are not yet officially recognized by the center.

This is not the place to comment on these choices, but it is noteworthy to observe that whereas the first photograph was the cause of a century of very interesting scientific studies, the carbon dating result, analyzed in Chapter 4, officially stopped them for a time that is even now undefined.

Between May 25 and 28, 1898, the lawyer Pia took the first photograph and discovered that the photographic negative (Fig. 2.15) showed a clearer rendition of the image, revealing with

Figure 2.15 First photographs of the face of the Man of the Shroud (in negative) taken by Secondo Pia in 1898 (to the left) and by Giovanni Sanna Solaro (to the right). The longer time of shutter speed and the stronger contrast in the right image highlight some details such as the hair.

high precision the features of a man and giving therefore a strong push to new studies also in the medicolegal field.

The photographic negativity featuring the body image was immediately read as a particular sign to be studied in detail, but the scholars who wanted to debunk the Shroud authenticity accused Pia of having forged the pictures. Even the evidence shown by the photographer Giovanni Sanna Solaro (Fig. 2.16) was not sufficient for weakening the controversy. It was necessary to wait 33 years before it could be shown that Pia was right; in fact, in 1931, during a public display, the Shroud was photographed again by Giuseppe Enrie [48], a professional photographer (Fig. 1.9).

In this book we will not detail all the conferences held in order to discuss the results obtained in the countless studies that have been carried out; we will only focus on the most significant moments of the twentieth-century scientific research, listing only the main symposia of this period.

The photographs that Enrie took in 1931 using a high-resolution orthochromatic film, to effectively enhance the contrast, allowed a

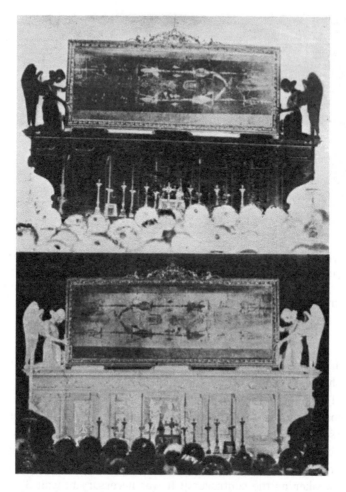

Figure 2.16 Giovanni Sanna Solaro, to demonstrate that the image of the Shroud is like a photographic negative, compared the negative picture (above) with the positive (below) of the altar above which the Shroud was placed in public displays.

detailed study of all the Shroud particulars and gave a remarkable pulse to scientific research, which resulted in two congresses:

- The First National Congress on Shroud Studies, Turin, 1939
- The International Sindonological Congress, Rome–Turin, 1950

Cardinal Michele Pellegrino, archbishop of Turin, nominated a commission [66, 67] that examined the Linen from June 16–18, 1969. In that occasion the first color photography was taken by Giovanni Battista Judica Cordiglia, a professional photographer who took also the first ultraviolet photographs. Some samples from the Shroud were also collected, among which were some blood scabs. Unfortunately, blood analysis was inconclusive, and valid results arrived only after 10 years.

Other conferences followed:

- The First US Conference of Research on the Shroud, Albuquerque 1977
- The United States Conference of Research on the Shroud of Turin, Amston 1978
- The Second International Symposium on the Shroud and public display, Turin, 1978

On November 23, 1973, the Shroud was exhibited for the first time ever on television, in color. On that occasion, there was a further examination of the Sheet, and botanist Max Frei, together with Gilbert Raes, a textile expert, was allowed to take some samples, among which was some surface dust.

The analysis made by the Shroud of Turin Research Project (STURP) in 1978, carried out in a methodical and detailed way, opened wide study and research opportunities in several scientific fields, and from that moment on, the most flourishing period for research on the Shroud started. This was also demonstrated by the proliferation of scientific conferences, which in this decade had an average frequency of 0.7 congresses per year.

On October 8, 1978, about 50 scientists from the STURP team spent 120 continuous hours conducting examinations on the Shroud in order to accomplish a multidisciplinary scientific investigation. The main purpose of this research was summarized in the following three questions:

(1) What is the Shroud body image composed of?
(2) How was the image produced?
(3) Is there blood on the Shroud?

Table 2.1   Continuous-hour initially estimated test plan

Tests	Number of persons	Expected time in hours	Notes
Photography (block 1)	8	6	4,5 hours: photomosaic
			1,5 hours: spectral coverage
X-ray radiography (block 1)	7	2	Preliminary tests of exposure times
X-ray fluorescence (block 1)	6	4	Feet blood
Infrared (block 1)	6	1	IR photo exposure tests
Spectroscopy (block 1)			The IR reflectance
Infrared (block 2)	10	7	follows the spectroscopy
Nastri (block 1)			(reflectance and UV fluorescence)
X-ray radiography (block 1)	7	7	Complete X-ray
(blocco 2)			radio-graphic coverage
Photography (block 2)	8	6	3.5 hours micro-photos
			2.5 hours spectral coverage
X-ray fluorescence (block 2)	6	10	Blood, face and fingers
Infrared (block 1)	6	2	Thermographic exposures
Infrared (block 1)	6	3	IR Photos

A test plan was set up for 48 continuous hours in front of the Shroud, reported in Table 2.1, but actually it was prolonged for 120 hours.

From these studies came out an official report in 1981, in which it has assumed that the Shroud enveloped the corpse of a man and that the body image has been imprinted on the Sheet in a way still unknown to science. Different hypotheses about the body image formation have been articulated, but it was observed that hypotheses reliable from the chemical point of view were unreliable from the physical point of view and vice versa.

The number of data collected from the American scientists, matched with the data acquired from some Italian scholars, gave an unlimited source of information described in several scientific publications. Whereas the American scientists preferred publishing their results in international scientific journals or in international conferences not properly dedicated to the Shroud, the Italian researchers chose the diocesan review *Sindon* or the symposia directly connected with the theme to publish their results.

The two most important and complete works that cleverly summarize the interesting results are those of L. Schwalbe and R. Rogers [129] and of E. Jumper, A. Adler, J. Jackson, S. Pellicori, J. Heller, and J. Druzik [85]; these two papers would have to be read by anyone who wants to start a scientific study on the Shroud.

Here are listed the conferences that followed:

- The 2nd National Congress, Bologna, 1981
- The New London Symposium, 1981
- The International Conference sponsored by IEEE, Seattle, 1982
- The 3rd National Congress, Trani (Apulia), 1985
- The Hong Kong Symposium, 1986
- The 4th National Congress, Syracuse (Sicily, Italy), 1987

On March 18, 1983, Humbert II of Savoy died, and as for his testamentary disposition, the Holy Shroud was donated to the Holy See in the person of the present Pope John Paul II. However, it is not yet clear if Humbert II had the authority to make such a gift, since Savoy's properties had been confiscated by the Italian State after World War II. As we can see, disputes over the Shroud are not over yet.

The growing scientific interest for the relic was abruptly interrupted in this period by the 1988 radiocarbon dating result, which officially dated the Shroud to the Middle Ages.

On April 21, 1988, samples of the Shroud were taken to undergo the radiocarbon 14 dating test. On the basis of this analysis the Shroud would date back to the Middle Ages, between A.D. 1260 and 1390, but as will be widely discussed in Chapter 4, the result obtained is not scientifically reliable.

This result moved all the scientists' attention to the analysis of this outcome and toward the reasons that could have caused it. It happened, therefore, that the research started to focus on the analysis of the radiocarbon dating method in detriment of the development of the scientific research on the general aspects of the Shroud.

In this period consistent proliferation of conferences led to an average frequency of more than 1.1 congresses per year.

- The International Shroud Symposium, Bologna, 1989
- The Shroud Symposium, Paris, 1989
- The 5th National Sindonological Congress, Cagliari, 1990
- The Symposium on the Shroud, St. Louis, Missouri, 1991
- The Shroud Symposium, Rome, 1993
- The Conference on the Shroud, Evansville, 1994
- The San Marino Conference, 1996
- The Esopus Conference, 1996
- The Nice Symposium, 1997
- The 3rd International Congress, Turin, 1998
- The Dallas Meeting of American Sindonology, 1998
- The Richmond Conference, 1999
- The Scientific Congress on the Holy Shroud of Turin, 2000
- The Orvieto Worldwide Congress "Sindone" 2000

In the night between April 11 and 12, 1997, a fire broke out in Turin's Guarini Chapel, quickly threatening the Shroud's bulletproof display case. Firemen saved the Shroud using a sledgehammer to break open the crystal case, and the Shroud was taken out from the cathedral (Fig. 2.17).

Two conferences followed:

- The 2nd International Shroud Conference, Dallas, 2001
- The 4th Symposium Scientifique International du Cielt, Paris, 2002

On June 20, 2002, after the Paris Symposium held in April, in which no mention was made of any program of restoration, a heavy intervention on the Shroud started and lasted till July 23: the patches sewn by the Poor Clare nuns and reinforced by Sebastian Valfrè were removed, and the fire-darkened shards still extending out from the weave into the burn holes were scraped off. A new backing cloth purchased 50 years ago has been sewn on the Shroud, and new scientific analyses have been carried out. This intervention has raised some perplexities both from the scientific and from the historic point of view.

In this process, also the Holland cloth sewn together with the patches was removed. It was also possible to obtain new interesting information on the relic, such as, for example, that referring to the

Figure 2.17 The Turin Cathedral and Guarini's Chapel under restoration after the 1997 fire.

back features of the Shroud after it was unsewn from the Holland cloth.

On this occasion Gian Carlo Durante took important photographs of the back of the Shroud; they show the presence of another body image, hidden away for centuries. The forensic science department of the Turin Police also took ultraviolet photographs of the back, but together with other data, this interesting information is still kept secret. Then, the Shroud has been placed in its new case in the Turin Cathedral (Fig. 2.18), the details of which are reported in Chapter 1.

Further photographs have been taken, and in 2008 the Italian company Haltadefinizione made ultrahigh-resolution photographs, which in 2013 were made available by an ad hoc app called Shroud 2.0.

Figure 2.18 The Shroud in its present reliquary. On the removable fabric that covers the case, the traditional Holy Shroud prayer: *TUAM SINDONEM VENERAMUR, DOMINE ET TUA RECOLIMUS PASSIONEM* (We revere Your Shroud, Lord, and [through it] we meditate on Your Passion) (Archdiocese of Turin).

Other analyses and surveys have been carried out, and many digital images have been acquired with the purpose to be useful for future scientific research, though until 2014 this material was not provided to the international scientific community.

In September 2002, Emanuela Marinelli created a mailing list through which a new group, nominated the Shroud Debate Project (SDP), started a scientific debate about some issues related to the Turin Relic. This group was composed of 10 or so Shroud scholars, mainly Italians and Americans, among whom was the first author.

After a full exchange of information, on the basis of the proposal of Mario Latendresse, E. Marinelli, R. Rogers, and the first author, the group disbanded and a new private mailing list group, named the Shroud Science Group on Yahoo! was founded and in 2013 counted more than 100 scholars.

From December 2002 and 2014 more than 25,000 messages have been exchanged, and among them, there are interesting scientific data on the body image, the blood rivulets, the pollen, and many other issues, still open.

Figure 2.19 Shroud Science Group logo.

The group is composed of academics from all over the world universities, researchers, and scholars and is characterized of a remarkable multidisciplinarity, which is very important for Shroud studies (Fig. 2.19).

In 2005 the group published a paper regarding the features of the Shroud, to be considered for the study of the body image formation. It also organized on August 14–17 the International Conference: The Shroud of Turin; Perspectives on a Multifaceted Enigma, Columbus, Ohio, 2008. Then, another conference followed, the International Workshop on the Scientific Approach to the Acheiropoietos Images, ENEA Research Center, Frascati, Italy, 2010.

The many others conferences and symposia on the Shroud cannot be listed in a few pages, but to have an idea, the reader can consider that many sindonologists personally held hundreds of public conferences on this topic.

This overview on the research carried out till today shows on the one hand the great scientific interest of the scholars for the Shroud and on the other hand the still lacking knowledge about the Sheet, especially as far as the body image formation is concerned.

This issue underlines the current limits of modern science and increasingly drives human knowledge to do more in-depth research in a field that seems to open new possibilities every time that someone tries to study a particular argument in detail; the work is still unimaginably wide, and only the future generations can reduce the big scientific gaps related to the most important relic of Christianity.

Chapter 3

Numismatic Investigation

What is the point of carrying out an analysis of the numismatic iconography of Christ rather than harking back to studies, also well developed, of His image depicted on icons, sculptures, and bas-reliefs? The answer is simple: Coins are easier to date back, even in very short periods of time; therefore it is easier to make a chronological reconstruction of the historic-iconographic path of the subject studied.

Furthermore, it is easier to highlight the changes of the typical depictions also according to the various historical happenings. For example, the effects of the iconoclastic controversy find direct confirmation in the different depictions of Christ in the Byzantine coinage.

The numismatic investigation in this chapter will introduce Christ representations on the coins according to the following development in order to make a more direct comparison with the Shroud image:

(1) First coins depicting Jesus Christ
(2) A selection of Byzantine and other coins depicting Jesus Christ
(3) A selection of Jesus Christ's portraits and a discussion of their details

The Shroud of Turin: First Century after Christ!
Giulio Fanti and Pierandrea Malfi
Copyright © 2015 Pan Stanford Publishing Pte. Ltd.
ISBN 978-981-4669-12-2 (Hardcover), 978-981-4669-13-9 (eBook)
www.panstanford.com

Each group of coins has a paragraph dedicated. It will be possible to analyze in a more-than-a-millennium time span, starting from the fifth century, the depictions of Christ on the Byzantine coinage, and in particular, the peculiar features of these representations will be compared with the Shroud face.

From the analysis of the historical developments of the different depictions of Christ and from some direct, also objective, comparisons with the Shroud face, it will be demonstrated, with reasonable certainty, that the Shroud was the reference image starting from the seventh century, not just for the Byzantine population.

This clearly contrasts with the result of the radiocarbon dating carried out in 1988, to which Chapter 4 is dedicated. As it is well known, the test outcome dated back the Shroud to the late Middle Ages (A.D. 1325 ± 65 years); but there will be the chance to discuss that the result obtained is actually improper, since, starting from the statistics analysis carried out on the basis of the data produced by labs, the resulting date of the Shroud is not reliable because of the presence of an environmental effect that involves a possible result variability of more than 1000 years.

So, not only can the Shroud easily exist in the seventh century at all, being a model for the coins minted with Christ portraits, but even more, on the basis of the results given by the alternative dating techniques developed in Chapters 6 and 7, a compatibility emerges between the dating interval obtained and the period of life of Jesus of Nazareth, that is, the first century A.D.

Obviously, the fact that the relic dated back to the first century A.D. does not automatically imply that it enveloped Jesus' body, since as seen in Chapter 1, from a strictly scientific point of view, the Man of the Shroud still remains with no name.

In numismatics there are specific technical terms to indicate particular coins. Referring to Byzantine coinage, it is useful to bear in mind the terminology reported in Table 3.1.

For the purposes of the following analysis, the *solidus* is one of the most interesting coins, together with the other gold coins, for the abundance of details that define the depicted faces. It was introduced by Constantine I in 309–310 and used throughout the Eastern Roman Empire till the tenth century. The *solidus* replaced the *aureus* as the main gold coin of the Roman Empire. The coin

Table 3.1   Some terms related to Byzantine coinage

Period	Gold coins	Silver coins	Bronze coins
First (310–approx. 700)	Solidus Semissis (1/2) Tremissis (1/3)	Hexagram (from 615)	Follis and others
Second (approx. 700–1092)	Solidus, Nomisma or Histamenon Tetarteron	Miliaresion (from 720)	Follis
Third (1092–approx. 1300)	Hyperpyron	Electrum or Aspron Trachy (gold alloy)	Billon Aspron Trachy Tetarteron and half Tetarteron
Forth (approx. 1300–approx. 1350)	Hyperpyron	Basilikon Assarion	Billon Aspron Trachy
Fifth (approx. 1350–1453)	//	Stavraton Half-stavraton Doukatopoulon	Tournesion Follar Aspron (Silver alloy)

weighed 1/72 of the Roman pound of pure gold, about 4.5 g (0.16 oz.) of gold per coin. Fractions of *solidus* were also produced, called *semissis* (half-*solidus*) and *tremissis* (a third of *solidus*). The word "soldier" is ultimately derived from *solidus*, referring to the *solidi* with which soldiers were paid.

Often, the mintage of a new solidus was directly authorized by the emperor and the coins depicting the face of Christ were considered by their possessors equal to a relic.

3.1 The First Coins of Christ

There are not so many in-depth numismatic studies [27, 104, 150] referring to the very vast coinage related to the depiction of Jesus Christ that were developed during the Byzantine Empire and that endured in the centuries after its fall. The lack of detailed information caused some controversies [106] about some Shroud scholars' assertions [151–153] that observed a remarkable similarity between the face of Christ depicted in the first millennium icons and the face of the Man of the Shroud.

For this reason, it is appropriate to explore when and how the numismatic representation of the Christ face developed throughout history in order to understand how the Shroud face influenced this representation. It is obvious that, once the Shroud's existence during the first millennium was verified, the still sustained hypothesis that the Shroud is a medieval artifact turns out to be even more unfounded.

Christianity became the official religion of the Roman Empire in A.D. 313; the Western Roman Empire fell in A.D. 476, whereas the Eastern Roman Empire, also known as the Byzantine Empire, survived for other 10 centuries, with Constantinople as its capital.

In agreement with Mario Moroni [103], in the Byzantine period, coins celebrated an event, thus making it official, so they have an irreplaceable value as a historical source. Furthermore, according to the Byzantine liturgical regulation, every image was represented with a rigorous identity concept that constituted itself *the reality illustrated*. By consequence, the images depicted had to portray the real features, avoiding every kind of inspiration of the engraver— hence explained the importance of the study of the details of the numismatic depictions: they have to be considered as historical evidence of truly happened events and not as an artist's fantasies.

The first images attributed [27] to Christ, discovered on the coins, date back to the half of the fifth century and are on one-of-a-kind-in-the-world ceremonial gold *solidi*. Two of them have been found, one related to the Licinia Eudoxia and Valentinian III wedding celebrated on October 29, 437, and the other commemorating the marriage between Saint Pulcheria and Marcian on July 28, 450; this last *solidus* is conserved at the Hunteriam Museum & Art Gallery of Glasgow University (Scotland).

On both coins, on the reverse, the standing figures of the emperor and the empress, united by the figure attributed to Christ, standing between them with His hand on their shoulders, and the writing FELICITER NUBTIIS (happy marriage) are depicted.

On these coins, the Christ face with short curly hair and beardless is typical of the first pictorial and iconographical images and is very similar to the Savior's representations in the Roman catacombs of the early centuries (see, for example, Fig. 2.6). The absence of elements that could make one hypothesize a similarity with the

Shroud face leads to the assumption that in those times the Christ image on the Shroud was not known in the Byzantine Empire yet.

After this coin, two more centuries had to pass before a mintage with Christ depiction appeared. In 692, canon n. 82 of the Council in Trullo (or Quinisextum) decreed that Christ had to be exhibited in images in human form instead of as the ancient lamb, as the custom was in that time, and simultaneously Emperor Justinian II Rinotmetus (685–695) minted the first official coins with the face of Christ: the gold *solidus*, *semissis*, and *tremissis* and the silver *hexagram* [22].

Breckenridge [27] observes that the Acts of the Quinisextum Council were never signed by the pope and that he refused to permit their publication within his domain: it is interesting to notice, confirming this historical record, the absence of any coins in this period with the face of Christ minted by the Italian mints that were under papal domain.

The first emperor minting a Christ effigy in Byzantine coinage was then Justinian II. During his first reign (692–695) he minted faces of Christ that can be grouped into two different types, the first type (in agreement with Breckenridge), reported in Fig. 3.1, and the second type, depicted in Fig. 3.35 (top right and bottom left), that are very similar to the Shroud face. During his second reign (705–711) the emperor minted Christ faces with Syrian-like features. In the next lines, and more in detail in the paragraph related to the face analysis, a description of these coins is provided.

On the coin (Fig. 3.1) the bust of Christ, in frontal position with a Cross behind His head, can be observed. He has long, wavy hair, a beard, and a mustache, wears a pallium[1] over a colobium,[2] bestowing a blessing with His right hand, and holds the Book of Gospels in His left hand. Written around His head is *Christ, King of those who rule*. On the reverse there is the emperor effigy with the sign *Lord Justinian, the servant of Christ*.

---

[1]A pallium is a liturgical vestment constituted by a band of white wool worn on the shoulder. By symbolizing the lost sheep that is found and carried on the Good Shepherd's shoulders, it signifies the bishop's pastoral role as the icon of Christ.

[2]A colobium is a tunic without sleeves or with short, close-fitting sleeves worn by the early monks.

Figure 3.1 Gold *solidus* of type I, the first period of Emperor Justinian II, minted in 692, depicting a Shroud-like face of Christ (coin kindly provided by US lawyer Mark Antonacci). The graph paper on the background allows one to determine the dimensions of this coin and of those in the following figures.

Although Emperor Justinian II was deposed in 695, between 692 and 695 there was a remarkable series of type II coin mintages depicting Christ, also attributed to several coin laboratories located in Constantinople and Sardinia [130].

If we recall Fig. 1.21, observing Fig. 3.1 we find ourselves in front of a Shroud-like face of Christ. Shroud resemblances of this face will be analyzed later; here it is interesting to observe that it is likely that not all the engravers had the chance to directly copy the face of Christ from the Shroud, and this is the reason why there is a considerable variety of interpretations caused by the reproduction of previous copies (Fig. 3.32).

The first engravers who could directly observe the relic did the most faithful depictions, whereas others just copied the previously minted coins or some then-available images painted from the Shroud, making the human features more similar to a perfect man but decreasing the particular characteristics that made the face pretty much singular, like the clear marks of His Passion, for example, the swelling on the cheeks caused by the suffered blows, the asymmetrical tear on the right side beard, and the asymmetrical hair shape weakened during the time.

As far as the direct observation of the Shroud by the engravers is concerned, some notes are worth considering. The hypothesis of Moroni is interesting: he reminds us that Jerusalem in 636 had been invaded by Muhammad's successor, and Sophronius, patriarch of Jerusalem, gathered the relics and had them carried to Constantinople; the Shroud could be among those relics, and hence explained the high grade of details with which the engravers reproduced the features of the Shroud face.

Another hypothesis is that the engravers were sent to Edessa, friend city of the Byzantine Empire, to copy the image conserved in that place. The Arabs, who occupied Edessa since 639, did not prevent the veneration of the Holy Face, also called *Mandylion* (the Shroud, according to several historians [154]), that was well known throughout all the East.

Furthermore, during the period of Emperor Justinian II, Byzantines were allowed to keep control on the Melkite community, which was constituted by iconophile Christians who collected a lot of relics. Byzantine artists had therefore access to those places, and among those relics, the Shroud could also have been conserved.

In addition to the indications of the Council in Trullo, according to the scholar Alan Whanger [150], the reasons explaining Christ's appearance on Justinian II coins were two.

The first reason refers to the fact that the empire was threatened by Muslims, who became military-aggressive, so the emperor needed a powerful palladium (a simulacrum believed to protect the Byzantine population) that could be used and recognized everywhere. Which palladium could be better than the Christ effigy minted on the highest-valued coins like the gold *solidus*?

The second reason is related to the fact that Egypt was the source of paper (papyrus) of the Byzantine Empire and it was controlled by Muslims. The emperor was irritated by the fact that Muslims had written a verse from the Koran on each of those papyrus sheets. So, as a reaction, the emperor decided to pay for those paper sheets with gold coins portraying the image of the face of Christ. This kind of reply had been considered by Muslims as offensive, so they melted those gold coins, using that metal for minting others with verses from the Koran.

Figure 3.2 Gold *solidus* dated back to the second period of Emperor Justinian II, 705–711, depicting the Syrian face of Christ, clearly different from the Shroud face.

Justinian II was overthrown following a coup d'état led by Leontios, and Monophysites, iconoclasts, took power; during the emperor of the usurper and of his successor coin mintages got back into tradition, without the face of Christ.

As soon as Justinian II regained the throne in the period between 705 and 711, he minted coins with Christ's face again (Fig. 3.2) but this time referring to a model different from the Shroud. Many hypotheses have been proposed to explain this change of reference image, but maybe the most probable is that in this case the emperor wanted to take the Camuliana Image as an example, which also should have been located in Constantinople at that time. The Syrian model of Christ's face depicted on these coins (Fig. 3.3) refers to a man with bushy curly hair and a short beard.

Since other face features like the big and close eyes are still similar to the Shroud image, it can be thought to a Shroud-like image concealed by purpose, maybe for reasons linked to the Byzantine iconoclasm that was already evident in that period.

After Justinian II, still because of the iconoclastic controversy, the habit of depicting the face of Christ on the coins disappeared for more than a century. Faces of emperors replaced it.

This iconoclastic struggle officially began with Leo III (717–741) and erupted from 725: depictions of holy images were forbidden

Figure 3.3 "Syrian" face of Christ, dated as post-sixth-century, burial crypt at Abu-Girgeh near Alexandria (from Breckenridge [27].

because they were considered as idols. By consequence, for about 150 years, the representation of Christ's face on Byzantine coins ceased.

According to Moroni [102, pp. 137–142], the Shroud disappeared from Constantinople in order to be protected during these struggles, and it was carried to Edessa in 754, where it was kept till 944, and then it was brought back to the Byzantine capital.

In 787 the Council of Nicaea established the clear difference between image veneration, allowed, and adoration, absolutely rejected, because only God can be adored. However, a second wave of iconoclastic controversies started and lasted till 843, when the March Synod finally revoked the iconoclastic decrees.

At the end of those struggles, the young emperor Michael III (856–867) led by his mother Theodora, iconophile, and by two other rulings, started again to mint gold coins with the Shroud-like face of

Figure 3.4 Gold *solidus* of Emperor Michael III, 856–867, depicting the Shroud-like face of Christ.

Christ (Fig. 3.4). It comes easy to think that also in this period some engravers would have had the chance to directly observe the Shroud before minting the coins.

3.2 Exploration of Other Depictions of Christ

From 869 to 944, years in which the Shroud was triumphantly brought back to Constantinople, there was a numismatic innovation: instead of minting just the Christ face, Byzantine emperors minted coins with the image of "Christ on the throne," in consistency with the Third Canon of the Fourth Council of Constantinople that imposed image veneration.[3]

The first emperor was Basil I the Macedonian (867–886), who introduced the Savior as a full figure with this writing: "IHS-XPS-REX REGNANTIUM" (Jesus Christ, King of those who rule), sat on the throne and nimbate. The image is that of a Christ Pantocrator, blessing, holding Gospels in His left hand. Other emperors followed his example, among whom were Leo VI (911–912); Alexander (912–913), Leo's brother; and Constantine VII (913–944) with Romanus I (920–944) and Romanus II (959–963) Figs. 3.5 and 3.6.

---

[3] From the third Canon: "We decree that the sacred image of our Lord Jesus Christ must be venerated with the same honor as is given the book of the Holy Gospels ... for what speech conveys in words, picture announces and brings out in colors."

Figure 3.5 Gold *solidus* of Emperor Basil I with his son Constantine (867–886) depicting Christ seated on the throne.

Figure 3.6 Gold *solidus* of Emperor Romanus I with Christopher (921–931) depicting Christ seated on the throne.

It must be observed that, in agreement with Breckenridge [27, p. 49], the attributes of the Christ figure are of the greatest importance in the imperial symbolism; therefore no details, even if at first sight can be considered noninfluential, have to be ignored. The detail of the enthroned Christ's right foot, smaller and almost right-angle rotated, is therefore noteworthy. This is consistent with the tradition of the lame Christ, already mentioned in Chapter 2, which probably derived from the interpretation of the particular position of the feet of the Man of the Shroud.

In agreement with the sindonologist Moroni [102, p. 3] the bare feet detail has not to be ignored. It is also depicted in the coins of Romanus I and Basil I in Fig. 3.5 and Fig. 3.6, and actually it does not suit high-rank people, but it is in close relation with what can be observed in correspondence to the dorsal body image of the Shroud, on which the bloody bare feet print is clear.

As previously mentioned, in 944 the Shroud was brought back to Constantinople; since then, mintages of the face of Christ reappeared, some of which are very similar to the Shroud face, but with more softened features compared to the first mintages of Justinian II.

An example in Fig. 3.7 shows a *histamenon nomisma* of Constantine VII Porphyrogenitus and Romanus (944), who ruled from 920 to 944 but probably minted this Shroud-like Christ effigy in 944. From then on, until the fall of the Byzantine Empire, several coins came in succession (from Fig. 3.8 to Fig. 3.15): the gold *histamenon nomisma* and *hyperpyron*, the *aspron trachy* in gold, silver and bronze alloy (among them the famous scyphate or cup-shaped coins), and the bronze *follis*.

Almost all these coins depict the face of Christ or Christ seated on the throne, both recalling the Man of the Shroud. Many of

Figure 3.7 Gold *histamenon nomisma* of Emperor Constantine VII and Romanus (944) depicting a blessing Christ very similar to the Man of the Shroud.

Figure 3.8 Gold *histamenon nomisma* of Basil II and Constantine VIII (976–1025) depicting a blessing Christ very similar to the Man of the Shroud.

Figure 3.9 Gold *histamenon nomisma* of Basil II and Constantine VIII (976–1025) depicting a second version of a blessing Christ very similar to the Man of the Shroud.

these[4] mark the Christ face with the initials IC XC, indicating Jesus Christ, in the Greek alphabet, "Ιησοῦς Χριστός" (Iesoús Christós),

---

[4]The emperors who minted these coins were Nicephoros II Phocas (A.D. 963–969), John I Tzimisces (A.D. 969–976), Basil II (A.D. 976–1025), Constantine VIII (A.D. 1025–1028), Romanus III (A.D. 1028–1034), Michael IV (A.D. 1034–1041), Zoe (A.D. 1041), Michael V (A.D. 1041–1042), Constantine IX (A.D. 1042–1055), Theodora (A.D. 1055–1056), Michael VI (A.D. 1056–1057A.D.), Isaac I Comnenus (A.D. 1057–1059), Constantine X (A.D. 1059–1067), Eudoxia (A.D. 1067), Romanus IV (A.D. 1067–1071), Michael VII (A.D. 1071–1078), Nicephoros III (A.D. 1078–1081), Alexios I (A.D. 1081–1118), John II Comnenus (A.D. 1118–1143), Manuel I Comnenus (A.D. 1143–1180), Andronikos I (A.D. 1183–1185), Isaac Comnenus (A.D.

Figure 3.10 Gold *histamenon nomisma* of Romanus III (1028–1034) depicting Christ seated on the throne.

Figure 3.11 Scyphate gold *histamenon nomisma* of Michael VII (1071–1078) depicting Christ seated on the throne.

considering the first and the last letter of each word ("Iς Xς"), written in capital letters.

The gold *histamenon nomisma* coin minted by Romanus IV Diogenes (1168–1171) depicts Christ (Fig. 3.14) in a different way, though very similar to the first coin dated 450. However, it has to be observed that the difference between the Christ face on the Marcian and Saint Pulcheria coin and the Christ face on the coin of

---

1184–1191), Isaac II (A.D. 1191-1195), Alexios III (A.D. 1195–1203), and Isaac II (A.D. 1203–1204).

Figure 3.12 Anonymous bronze *Follis* ascribed to Romanus IV (1068–1071) depicting a Shroud-like face of Christ and on the reverse God's mother.

Figure 3.13 Scyphate bronze *aspron trachy* of Alexios Comnenus (1143–1180) depicting Christ seated on the throne.

Romanus IV is that the first represents a beardless man with short hair; instead in the last the face is sindonic, with a beard and long wavy hair, showing that the Shroud image was the predominant model in that age.

With Emperor Romanus III (1028–1034) the mintage of Christ seated on the throne was resumed after about 90 years of absence

Figure 3.14 Scyphate gold *histamenon nomisma* of Romanus IV Diogenes (1168–1171) depicting Christ. On the left is Christ crowning Romanus IV (on the left side) and Eudoxia (on the right side); on the right, Michael, Constantine, and Andronicus; and on the bottom, detail of the Shroud-like face of Christ with long hair and a beard and the identifying writing "Iς Xς".

and this lasted till the empire's fall with Alexios III (1195–1204), in 1204, after the siege of the Fourth Crusade.

In addition to Emperor Justinian II in his second period of reign chose a Christ face depiction different from the Shroud face, Emperor Manuel I Comnenus (1143–1180), too, preferred to refer again to a younger and beardless Christ, the Christ Emmanuel (Fig. 3.15). These two exceptions compared with the countless sindonic images, confirming the real importance that the Shroud should have had in those centuries.

Figure 3.15 Scyphate gold *hyperpyron*, Manuel I Comnenus (1143–1180) depicting a young and beardless Christ Emmanuel.

Hereinafter the different details of the faces of Christ imprinted on the Byzantine coins will be compared with the Shroud face, but it is important to observe as it is that Byzantine iconography in general defined severe and detailed canons to which all the artists had to submit if they wanted to reproduce icons and therefore also the face of Christ.

As it will be shown, after the analysis of the Byzantine influence in the world, the predominant canon of the Christ depiction on the coins directly harks back to the Shroud image.

3.3 Byzantine Influence in the World

With the fall of the Byzantine Empire it could seem logical that also the depictions of Christ on coins would have ended. Instead, the tradition of portraying the effigy of Christ's face and His depiction seated on the throne endured during the centuries till the present day. It is interesting to observe that, over the course of centuries, many of the typical details of the Byzantine canons were gradually abandoned, and the depictions of Christ faces were less and less comparable with the Shroud face.

Also before the empire's fall in 1204, there were some coins with the Christ effigy minted by Byzantine's friend states or countries

Figure 3.16 Scyphate silver *ducat* of Ruggero II (Palermo 1130–1140) depicting Christ's face.

Figure 3.17 *Follis* of Ruggero II (Messina 1105–1154) depicting Christ's face. On the reverse, Arabian influences can be noted.

that came in touch with their culture. For example, Ruggero II (1130–1154), king of Sicily (Italy), who for political reasons had relationships with the Byzantine Empire, minted, in the cities of Palermo (Fig. 3.16) and Messina (Fig. 3.17), coins with the Shroud-like face of Christ, even if they do not reach the details represented in the coins observed till now.

There are also proofs that even Russia and Denmark had relationships with the Byzantine Empire already in the tenth

Figure 3.18 Danish *penny* by King Sweyn II Estridsson Ulfsson (1130–1140) depicting Christ seated on the throne.

century. Czar Vladimir the Great, converted to Christian religion, also used coins with the face of Christ to evangelize Kiev. In the ancient Danish city of Lund coins minted depicting an enthroned Christ recalled the Byzantine type (Fig. 3.18).

The tradition of minting images of Christ on coins, after the end of the Byzantine Empire, seems to have spread. First of all, the rulers who came in succession after the fall of the empire kept on minting the effigy of Christ, though respecting in a less marked way the typical rigor of the previous reproductions; in fact they evolve into a less refined depiction with a more free interpretation.

On the other hand, then, knowing that the Shroud had been purloined in Europe after the sacking of Constantinople in 1204, it is obvious to think that the degrading of Christ depictions was also due to the fact that there was no possibility anymore to reproduce images observing the Shroud directly. An example of this is the silver coin (*half-hyperpyron*) minted by Manuel II Palaeologus in 1391–1423 (Fig. 3.19).

It is interesting to observe the rare coin minted by Joscelin I de Courtenay, Count of Edessa, between 1119 and 1131 (Fig. 3.20) depicting the bust of Christ, because it shows how still active was the cult for the image of the face of Christ in that place, in memory of the Shroud there exposed in the past.

Figure 3.19 Silver coin (*half-hyperpyron*) minted by Manuel II Palaeologus in 1391–1423, depicting the bust of Christ.

It is known that the Republic of Venice kept tight commercial relationships with Constantinople for a long time, and by consequence, it acquired its styles and habits. Early before the fall of the Byzantine Empire, in 1202, the republic, under Doge Enrico Dandolo (1192–1205), being inspired from the Byzantine type of Christ coinage, started a series of mintages of Christ that lasted for several centuries till the fall of the maritime republic. An example is the depiction of

Figure 3.20 This rare coin minted by Joscelin I de Courtenay, Count of Edessa, between 1119 and 1131 depicting the bust of Christ, shows how still active was the cult for the image of the face of Christ in that place, in memory of the Shroud there exposed in the past. Unfortunately, the signs of the time do not allow one to observe face details.

Figure 3.21 Venetian *grosso* minted by Enrico Dandalo (1192–1205) depicting Christ seated on the throne with clear Byzantine features.

Christ seated on the throne in Fig. 3.21 that portrays Him according to the Byzantine canons, even with the right-angle-rotated right foot. It is also evident how the features of the face with long and asymmetrical hair are typical of the Byzantine coins as well as of the Shroud.

The Venetians, however, were not satisfied just by copying the typical Byzantine depictions of Christ, but they went further beyond minting also different representations, always being coherent with the canons acquired through the observation of the Shroud image. An example (Fig. 3.22) is the *mezzanino* minted by Doge Andrea Dandolo (1343–1354) depicting a resurrected Christ with asymmetrical long hair and a bipartite beard.

The gold *sequin* minted by the Venetians, for example, the one by Francesco Foscari (1423–1457) in Fig. 3.23, shows a new image of Christ within an oval or "*vesica piscis*" (fish bladder), called "*Mandorla*," Italian for almond, that is, an oval frame with decorative function enclosing a sacred figure therein highlighted. It is a Romanesque-Gothic decorative element.

The *Mandorla* has a double value: Both referring to the almond fruit and to the seed in general, it is a clear symbol of life, so an attribute for Who that is *the way, the truth and the life*; and as intersection of two circles it represents the communication

Figure 3.22 Venetian *mezzanino* (T/N: half-*grosso* coin) minted by Andrea Dandolo (1343–1354) depicting Christ resurrected with the Shroud face.

Figure 3.23 Venetian *sequin* minted by Doge Francesco Foscari (1423–1457) depicting resurrected Christ within a Mandorla.

between the two worlds, the material level and the spiritual level, the human and the divine. Jesus, made man, becomes the only mediator between the two realities; therefore He is depicted within the intersection.

Again, the features of the Christ face in these coins are sindonic and even the right foot is often portrayed as smaller and rotated as in the Byzantine tradition. The *sequin* must have to be successful as the gold coin since it was adopted also in other Italian states of that

Figure 3.24 *Sequin* minted by the Roman Senate (1390–1410) depicting Christ within a Mandorla.

age and it was also minted for the Crusaders. As examples, see the *sequins* minted by the Roman Senate in 1390–1410 (Fig. 3.24), by the Neapolitan Roberto d'Angiò for the Maona of Chios (medieval company of investors) (1346–1364), and by the Milanese Filippo Maria Visconti (1421–1436).

Also the Venetian *lira*, for example, that minted by Pietro Mocenigo (1474–1476) in Fig. 3.25, displays a new depiction of a resurrected Christ, always following the Byzantine canons, and therefore the Shroud features, of a Jesus with long hair, beard, and mustache.

In Rome, the Senate, and also some popes, around 1300, minted other coins with the face of Christ, perhaps inspired instead by the Holy Face that should have been conserved in the city at that age. Also this depiction does not have remarkable differences from the Shroud face, but the features of the coin in Fig. 3.26, typical of those years, and not very refined, do not allow to make the right comparisons.

Arising from the Venetian numismatic depictions, Stefan Uros I (1243–1272) from Serbia minted coins, the *dinars matapan*, very similar to the *grossi* depicting Christ seated on the throne, also with the rotated right foot anomaly (Fig. 3.27). The face takes inspiration from the Byzantine canons as well.

Figure 3.25 Venetian *lira* minted by Doge Pietro Mocenigo (1474–1476) depicting a resurrected Christ.

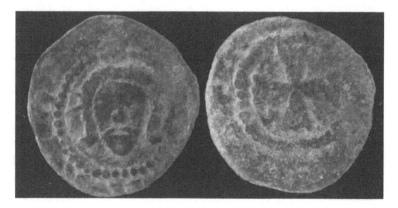

Figure 3.26 *Picciolo* minted by the Roman Senate (about 1300) depicting the face of Christ.

Also Dubrovnik (Italian name Ragusa), that became part of the Republic of Balkans between 1284 and 1372, minted coins with Christ within an oval, as depicted on the famous gold *sequins*. The features are still in compliance with the Byzantine canons (Fig. 3.28).

Similarly, the Armenian state of Georgia, under the reign of Rusudan (1223–1245), and the area of Anatolia, Northern Mesopotamia, under the reign of Fakhr al-Din Qara (Kara) Arslan (1144–1174) from the Artqid dynasty, minted coins with the Byzantine type of Christ face.

Figure 3.27 Serbian silver *grosso* (dinar) minted by Stefan Uros I (1243–1272) depicting Christ seated on the throne like in the Venetian *grossi*.

Figure 3.28 Silver *grosso* (*dinar*) from the Republic of Balkans minted by anonymous (1284–1372) depicts Christ within an oval (*Mandorla*) harking back to the famous Venetian *sequins*.

Finally, also Bulgaria shows some Byzantine influences, since it minted coins with less refined depictions of Christ but vaguely consistent with the above-mentioned canons anyway. An example is the silver *grosso* by Theodore Svetoslav (1300–1322) depicting Christ seated on the throne (Fig. 3.29). The anomaly of the rotated right foot is also carved on this coin made with more rough features: this peculiarity, therefore, had to belong to a canon carefully observed by all the engravers.

Figure 3.29 Bulgarian silver *grosso* (dinar) of Theodore Svetoslav (1300–1322) depicting Christ seated on the throne.

The fact that such peculiar features referring to the human figure of Jesus Christ did persist during the centuries and that they were codified in canons that strongly influenced both the numismatic iconography and the painting iconography in general makes one think of a very important and universally recognized reference model.

The following observation has to be carefully borne in mind: in the numismatic depictions of Christ, and not only these, some peculiar features persist, such as the slanted foot, the asymmetrical beard and hair, the swollen, because tumid, cheeks, etc. These characteristics, that obviously disfigure the beauty of the face and of the human figure, are in strong contrast with the classic trend of representing Jesus Christ as the most handsome man.[5]

There are not only coins depicting Christ, but also some interesting lead seals displaying the face of Christ and that have some references to the body image of the Shroud. For example, the Bulgarian seal depicting a blessing Christ, dated back to the time of Czar Peter I and Maria Lecapena (927/928), is displayed in Fig. 3.30. Even in this case the representation is less refined than the Byzantine, but nevertheless, some Shroud features still appear, like the face asymmetry and the hair, longer on the left side.

---

[5] Refer to Psalm 45:2 : *You are the most handsome of the sons of men.*

Figure 3.30 Bulgarian lead seal of Czar Peter I and Maria Lecapena (927/928) depicting a blessing Christ face.

These characteristics, beyond making us wonder on the reason why the engravers made the figure of Christ ugly, indicate that the reference model was one, and very peculiar, with evident flaws and asymmetries. Why, then, cannot we see a direct allusion to the Shroud image that exactly contains these features? The next paragraph will debate a study of the Shroud face details reproduced on the coins in order to confirm this close relation of images.

Reviewing what has been introduced till now, we can sum up:

- The precision of the Byzantine golden mintage is very useful for the comparison of several details that are in common with the Shroud image.
- The first coins depicting Jesus Christ seem to date back to 437–450 and celebrate the wedding between Valentinian III and Licinia Eudoxia and between Marcian and Saint Pulcheria, but the face is that of a generic, not sindonic, beardless man with short hair.
- There are no other mintages depicting Jesus Christ until 692, a date in which the canon n. 82 of the Council in Trullo decreed that Christ would be represented as a man and no more as a lamb. In the same year, Justinian II minted a series of coins depicting the bust of Christ with Shroud-like characteristics.

- The iconoclast struggles forbid Christ representations up to period between 856 and 867, years in which Michael III resumed the mintages of the face of a sindonic Christ.
- From 869 on, with Basil I, the Byzantine emperors minted coins depicting not only Christ's face but also a full-figured Christ seated on the throne. On these coins there are other Shroud features, like that of the rotated right foot.
- Christ's depiction on coins has not been a phenomenon just limited to the Byzantine area, but it spread in different countries throughout the world, even after the empire fell in 1204.

3.4 Coin Analysis of the Faces of Christ

A brief observation of Christ images, depicted in Byzantine coins at the beginning of the numismatic dissertation, highlights a direct reference to a well-defined canon.

This has also shown by the fact that in the cases in which the face of Christ made reference to other images, like in the case of Marcian and Saint Pulcheria, Justinian II (Period II), and Manuel II Comnenus, the difference among the minted faces is clearly evident. Below, the characteristics linking the Byzantine face of Christ to the Shroud face are shown in more detail.

The first photograph taken by Secondo Pia allowed a more direct access to the Shroud. In the first years of the twentieth century P. Vignon [145] and later other scholars in the world made a comparison between the image of the sindonic face of Christ and the various images of the iconography. Vignon defined some distinctive markings of the Shroud image that are common in numerous iconographic depictions [115].

(1) Two wisps of hair in the center of the forehead
(2) Hair on one side of the head longer than on the other side
(3) A U shape between the eyebrows, often square-bottomed
(4) A downward V-shaped pointing triangle on the bridge of the nose

(5) One raised eyebrow
(6) Large, round eyes
(7) Accents on both cheeks, somewhat lower on the right cheek
(8) A forked beard
(9) A gap in the beard below the lower lip
(10) An enlarged left nostril
(11) An accent line below the nose
(12) A dark line just below the lower lip

Not everybody agrees in recognizing all these markings, stating that some among them, like, for example, the U shape between the eyebrows, are more subjective and therefore less meaningful. For this reason, also for the purposes of the numismatic analysis of the Shroud face that will be developed below, only the more evident markings are taken into consideration.

(1) A lock of hair at the center of the forehead, referring to the reversed "3"-shaped forehead wound (which, as shown shortly afterward, could not be represented like that).
(2) Wavy, shoulder-length hair longer on the left side, similar to "payot," the side curls of Orthodox Jews. The right side of the hair is less bushy, maybe because it is partially torn, like the beard.
(3) A high-arched left eyebrow due to a hematoma.
(4) Protruding cheekbones, the right more than the left one, because of the swelling caused by the blows.
(5) A long bipartite beard; the right side is quite sparse in some areas, probably because it is torn like the hair.
(6) A gap in the beard below the lower lip.
(7) Broken nasal cartilage; therefore the nose is slightly flattened and turned to the right side.
(8) A protruding lower lip.

Let us take a careful look to the peculiar lock of hair on the forehead, typical of almost all mintages of the Christ face. This particular derives almost certainly from the famous reversed "3"-shaped forehead mark caused by a blood rivulet (Fig. 1.15). Since canon n. 82 of the Council in Trullo (692) dictated that the image of Christ could not be depicted with the marks of the Passion, it is easy to think that the engravers, to equally reproduce that particular blood mark typical of the Shroud image, changed it to a lock of hair.

Finally, another mark that would confirm the congruence with the Shroud face is that related to some rare coins that seem to show the face of Christ with His eyes closed. However, this little detail is often not so clear to be able to affirm with certainty that the eyes are actually shut, leaving out some possible mintage imperfections.

A discover of interest is that of a coin minted by Justinian II between 692 and 695, the *semissis* (half of a *solidus*) in Fig. 3.31 that seems to show more clearly the face of Christ with closed eyes. To verify this detail it can be useful to make a comparison with the

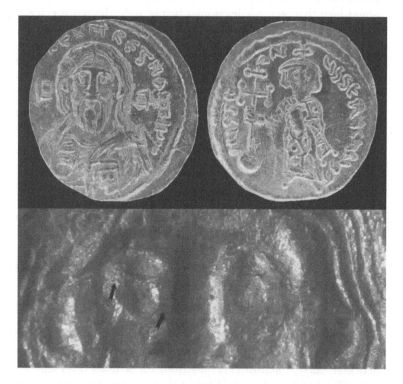

Figure 3.31 On the top, *semissis* of Justinian II (692–695) depicting a particular of the face of Christ, type II, period I. On the bottom, the detail of the eyes apparently closed that harks back to the Shroud face. It seems that two tears are also portrayed around the right eye, indicated by the arrows.

face of the coin on the top left in Fig. 3.33 where the iris is well highlighted in both eyes.

Closed eyes would seem a clear sign of the fact that the engraver followed the image of the Shroud face to reproduce this detail and that he would never have done it if he had had the intention to depict a normal face of live Christ.

This coin would definitely confirm that the Turin relic was taken as the model of the face of Christ in the coins minted by Justinian II and would lead to the thought that the engravers that observed the Shroud were attracted by some face details and decided to reproduce those that impressed them more.

The coin in Fig. 3.31 seems perhaps to highlight also two tears below the right eye. If it actually was like that, it could be also thought that the engraver wanted to reproduce the tear of the right eye that today is not really evident on the Shroud image anymore but at that time it was easier to be observed. This is why, as is well known, the contrast of the body image on the Shroud becomes less intense through time, and 14 centuries ago it was obviously easier to distinguish the tear particular. See also the face in Fig. 3.35 (bottom left) as further confirmation: it seems to show two tears, highlighted by the arrows.

Also nowadays, it seems that the scholar Aldo Guerreschi succeeded in distinguishing the tears detail after enhancing by artificial means the contrast of a photograph of the Shroud [74]. Then, this coin could also confirm the study of Guerreschi.

It has been shown (Fig. 3.2) that starting from 692 Emperor Justinian II minted the face of Christ on the coins, and it is easy to think that the first coins had been minted by engravers who had the possibility to observe in person the Shroud face in order to use it as a reference model. This deduction derives from the fact that on these coins there are details that no artist could have reproduced without having the Shroud as a sample of reference.

As the years went by, till 695, other engravers of other coin laboratories in the empire reproduced the face depicted on the first coins, probably without having the possibility of observing the original; therefore they lost some of the peculiar details of the Shroud face, producing a more refined and common portrait with less evident marks of the Passion. Breckenridge [27, p. 20] notices

Figure 3.32 Set of iconographical representations of the face of Christ in the various mintages of a gold *solidus* by Emperor Justinian II, 692. The first column on the left displays the coins classified as type I; on the right those of type II. The grade of similarity with the Shroud face seems to become less intense from the top left to the bottom right.

that, in the west of the Byzantine Empire, a certain amount of greater liberty in creating variant types was exercised in following the established criteria, and this can explain the remarkable variety of the existing mintages.

Figure 3.32 highlights the variety of the faces of Christ minted on gold coins only between 692 and 695 in the empire of Justinian II. It can be observed, from the top left (type I) to the bottom right (type II), how the typical features of the Shroud-like image fade.

The face of Christ depicted on one of the first mintages of Justinian coinage (type I) is that displayed in Fig. 3.33, in which many Shroud markings are evident; the numbers and the corresponding black lines in the figure underline some remarkable similarities between the two faces.

(1) Wavy hair and asymmetry between the right side (shorter) and the left side (longer)
(2) High-arched left eyebrow due to a hematoma
(3) Reversed "3"-shaped forehead wound interpreted as a lock of hair at the center of the forehead
(4) Very close, large, round eyes
(5) Contusion next to the right eye
(6) Protruding cheekbones because of swelling

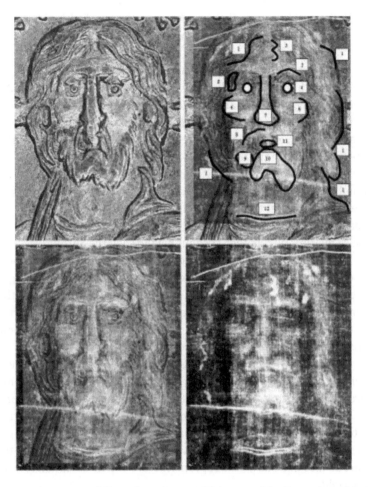

Figure 3.33 Face of Christ (type I, period I) in one of the first coins minted by Emperor Justinian II, probably in 692 (top left) compared to the Shroud face (bottom right). The other two images are given by the overlapping of the two faces.

   (7) Flattened and slightly curved nose
   (8) Long mustache
   (9) Sparse beard on the right side due to tear
 (10) Bipartite and asymmetrical beard
 (11) A gap in the beard below the lower lip

(12) Same shape of the wrinkle on the neck (double-lined)
   interpreted as the edge of the dress

In support of the thesis of the correspondence between the
two faces there are also the more than 100 points of congruence
determined by the scholar Whanger [151–153] on a similar coin; it is
not the case to go on with the analysis of other details to understand
that the engraver should have seen the Shroud before performing
the face shown in the figure.

Someone could look to "coincidence" in order to justify the
number of similarities, but making a rough probability calculus,[6]
we can affirm that this "lucky" engraver would have had only seven
chances in one billion of billions of different possibilities of hitting
these features all together without having seen the Shroud.

To better understand the meaning of this result "seven chances
in one billion of billions" we can make a similar example referring to
the roulette game. It would be much easier to hit for 10 consecutive
times the number 36 than an engraver depicting the face reported
in the figure without having seen the Shroud.[7] Obviously, this is not
absolutely impossible but extremely improbable.

As just shown, the coins minted by Justinian II after 692 show a
general attenuation of the Shroud markings, but this reduction is not
progressive during the centuries and there are also some exceptions.

For example, some rare coins minted by Justinian II in the period
after 692 display an interesting feature that was not observed by the
first engravers: the nose twisted to the right in consistence with the
Shroud face that shows the septum fracture and its deviation toward
the right.

Observing the Shroud's typical feature of the twisted nose, some
American scholars [82], against the Shroud's authenticity, criticized

---

[6] The details of this calculation, in which peculiar notions and theorems of probability
calculus are used, can be found in Section A.3 of the appendix. Therefore the reader
can refer to it.

[7] In the roulette game the numbers are 37. As the chance of hitting a single number is
1 on 37, namely $\frac{1}{37}$, the chances of hitting for 11 consecutive times the same number
at this game are $\left(\frac{1}{37}\right)^{11} = \frac{1}{37^{11}} = \frac{1}{1.78 \cdot 10^{17}} = 5.62 \cdot 10^{-18}$, those of hitting the same
number for 10 consecutive times are $\left(\frac{1}{37}\right)^{10} = \frac{1}{37^{10}} = \frac{1}{4.81 \cdot 10^{15}} = 2.08 \cdot 10^{-16}$,
whereas the odds calculated for the artist are $7.26 \cdot 10^{-18}$.

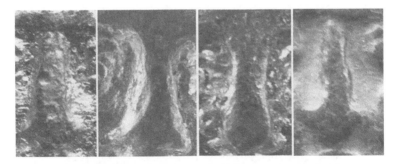

Figure 3.34 Details of the nose of the Christ face in some gold coins: from left to right two different mintages by Justinian II display the nose twisted to the left and a more common straight nose, respectively. On the right a nose with a more complex twist is displayed, much more similar to the Shroud face, in a mintage by Basil II and Constantine VIII (976–1025).

it, affirming also that a unique case could have been due to a minting flaw or to a blow on the coin during its circulation.

First of all it has to be observed that this feature is not unique in its genre, because other coins with the particular of the twisted nose have been found, even from different mintages. Second, observing the particular of the twisted nose magnified, an engraving continuity can be noticed that cannot be attributable to blows.

For example, Fig. 3.34 compares a common straight nose (in the center) with the twisted nose of the coin of Justinian II and with another skewed nose (with analogous features of the previous) of a coin minted by Basil II and Constantine VIII in 976–1025, which perhaps better reproduces the Shroud profile with a broken cartilage.

It is interesting to observe the extreme accuracy of the production and the skill of the engraver, which succeeded in reproducing details on the order of one-tenth of a millimeter (0.004 in.) without using optic magnifying means. It has to be remembered that the dimensions of the nose are about 2 mm (0.079 in.).

The face of the *solidus* of Michael III (856–867), instead, shows a flaw on the nose tip (Fig. 3.35) that, observed magnified, clearly turns out to be a blow taken during its circulation.

The features of this face of Christ reveal a further attenuation of the Shroud characteristics, though the general structure is still

Figure 3.35 Christ faces. On the top left one of the first mintages of Justinian II (type I, period I, dating 692). On the top right a mintage of Justinian II (type II, period I) in which the figure is a little polished from the signs of the Passion but discloses an additional peculiar Shroud feature: the right twisted nose. On the bottom left, to make a comparison, a mintage of Justinian II (type II, period I) that shows a straight nose but seems to reproduce two tears, highlighted by the arrows. On the bottom right, a mintage of Michael III (856–867) in which some markings, even if present, are softened. Picture by G. Concheri, University of Padua.

similar: asymmetrical long hair, close eyes, mustache and beard, and typical lock of hair in the center of the forehead.

The first coins minted by Theodora, iconophile, with his son Michael III, depict the face of Christ more similar to the Shroud than that minted after by Emperor Michael. This leads to the thought that

Figure 3.36   Christ face of the gold *solidus* of Emperor Basil I (867–886) on the left and on the right a solidus of Romanus I with Christopher (921–931).

the face of these last coins had been copied from the face of the first, without the possibility of observing the original image.

Also the face of Christ seated on the throne in the coins of Emperor Basil I (867–886) and of his successors reproduces the features of the coins minted by Michael III. Two examples are displayed in Fig. 3.36. The arrival of the Shroud in Constantinople in 944 defines a moment of renewal also from the numismatic point of view and the faces get back to the Shroud-like features.

However, these faces do not display the roughness of the first depictions and the result is a face of Christ less marked by the signs of the Passion and more handsome.

It is interesting to observe that, though the features are very soft, some Shroud characteristics are still evident, especially as far as the beard and mustache are concerned (Fig. 3.37).

Obviously, the gold coins are more interesting because they are more accurate; therefore the face of Christ is more detailed. The different coins, however, display depictions of Christ linked to the established criteria in different ways, and in some cases, some freedom of choice due to engravers can be noticed that perhaps did not have the chance to directly see the Shroud.

Probably, then, some engravers who minted Byzantine coins in cities far from Constantinople, like, for example, Ravenna and Carthage, felt free to redefine the subject just sketching some canons. It is the case, for example, of the gold coins displayed in Fig. 3.38.

Figure 3.37 On the left faces of Christ in the gold coins of Constantine VII and Romanus (945) and on the right those of Basil II and Constantine VIII (976–1025). The hair wave on the left is a very recurring feature, such as the longer, wavy hair on the left.

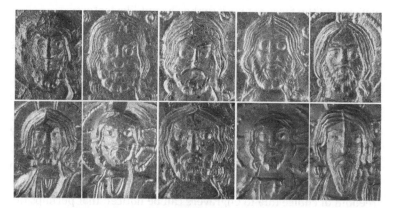

Figure 3.38 Various faces of Christ in gold Byzantine coins minted from the tenth and twelfth centuries. Even though the features are less in accordance with the Shroud, some typical markings are well evident, among those, the long, wavy, asymmetrical hair, swelling of the cheekbones, and a lack of beard under the lower lip.

Figure 3.39 Various gold coins of *histamenon nomisma* minted by Basil II and Constantine VIII (976–1025). It is interesting to observe the variation of Christ's face that is more Shroud-like on the left.

It must also be remembered that, during relatively long periods of reign, like the 50 years of Basil II (from 976 to 1025), the possibility to directly observe the Shroud by part of the engravers, commissioned by the emperor to reproduce the face of Christ, was quite variable and in some cases they reproduced copies of copies.

Figure 3.39 shows an example. The 10 gold coins of *histamenon nomisma* show how Christ's face varied along this long period of Basil II. While the coins on the left of the figure represent different interpretations of a TS-like face, those on the right have lost some general feature of that face, while some typical details probably imposed by the Byzantine canons like the asymmetrical beard or the tuft of hair (remembering the reverse "3" of blood) are still evident.

As the Shroud was triumphantly taken in Constantinople in 944, it is easy to think that the coins at the left of Fig. 3.39, which reproduce a TS-like face, were those minted in the first years of the reign (about in 976), while those on the right were minted after. This observation can be perhaps used from a numismatic point of view to codify the dating method used at these epoch that seems based on the different signs (points, lines, and circles) posed in correspondence of the nimbus.

In the period from 945 to the fall of the empire in 1204, the majority of bronze mintages of the face of Christ in Byzantine coins does not display a high number of Shroud-like characteristics, also because the features are more rough than the gold coins. However,

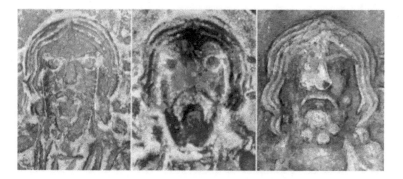

Figure 3.40 Faces of Christ in the bronze coins (*follis*). On the left and in the center those minted by Nicephorus III (1078–1081); on the right those of Basil II and Constantine VIII (1020–1028). Picture by G. Concheri, University of Padua.

some more rare coins in less precious alloys than gold highlight interesting details, less underlined by other engravers.

It is maybe the case of some artists that, having rubbed in their mind the Shroud face after observing it closely, wanted to immortalize some additional particularities.

For example, the two bronze *billon aspron trachy* in Fig. 3.40 of Basil II with Constantine VIII (1020–1028) and that of Nicephorus III (1078–1081) linger on the detail of the asymmetrical and sparser on the right beard. In the center the beard is asymmetrical and bipartite. On the left, instead, the artist wanted to highlight the less bushy beard on the right side by a lack of metal. On the right, instead, the artist, having at his disposal rough tools, simulated the beard using some disks but, to underline the lack of beard on the right, he placed these disks asymmetrically, leaving an evident blank.

The faces of Christ minted in the twelfth century in the reign of Sicily (Ruggero II and Guglielmo I) in Fig. 3.41 on the left and in the center hark back to the faces of Byzantine coins but are less rich in particulars and do not pay specific attention to the canons. Notwithstanding, a reference to the Shroud face still remains, especially for the beard and mustache and for the configuration of the shoulder-length hair, although no more asymmetric.

In the Republic of Dubrovnik (Republic of Ragusa), in the Balkans, about the fourteenth century, the face still keeps Shroud-

Figure 3.41 Faces of Christ on Sicilian and Dalmatian coins: starting from the left, silver scyphate ducal minted in 1140 by Ruggero II in memory of his ordination as Duke of Apulia; scyphate ducal of Guglielmo I (1154–1166); and Republic of Dubrovnik (1284–1372) with a more Shroud-like face.

like features, even though the mintage is less refined (Fig. 3.41 on the right). The hair is less wavy, but the bipartite beard appears again. Also the pronounced cheekbones make reference to the Shroud swellings.

As already shown, starting from 1202, with Doge Enrico Dandolo (1107–1205), the Republic of Venice minted silver *ducat* depicting Christ seated on the throne with typical Byzantine features. Such coins, then called *grosso matapan*, had great success in the Mediterranean area for several centuries [1].

The face of Christ on these coins is less similar to the Shroud face, but it still has some of its features. It is easy therefore to think that the Venetians did not have the chance to directly observe the Shroud but that they took the cues from its body image considering the Byzantine canons that probably described the face features in details and the figure characteristics more in general.

The face becomes more symmetrical and round shaped; the marks of the Passion, like the swollen cheekbones, are less evident; but some features are maintained anyway, like big eyes, beard, mustache, and long, often asymmetrical, hair (consistent with the Byzantine tradition, the left side of the hair is longer over the shoulder).

It is noteworthy that the Republic of Venice minted other coins depicting Jesus Christ. Among them, the silver *mezzanino*, minted by Andrea Dandolo (1343–1354), display a face with features different

Figure 3.42 Faces of Christ on Venetian coins: starting from the top left, silver *grossi* by Jacopo Tiepolo (1229), Lorenzo Tiepolo, Jacopo Contarini and Giovanni Dandolo, and Francesco Foscarini till 1457; on the bottom right, *mezzanino* by Andrea Dandolo (1343–1354).

from those of the *grossi*, but though dimensions are of just a few millimeters, they have Shroud-like characteristics, especially for the hair shape (Fig. 3.42).

Also other coins, among which the gold *sequins*, depict Christ, but in this case the face is far and far away from the Byzantine canons, though maintaining the general features (Fig. 3.43).

Besides these Italian states, also Serbia (Fig. 3.44) and Bulgaria minted coins with Christ's face, following the Byzantine inspiration

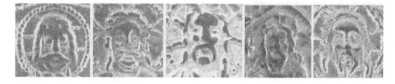

Figure 3.43 Faces of Christ on gold *sequins*: starting from the left, coins minted by Roberto d'Angiò for the Maona of Chios (1346–1364), by the Roman Senate (1390–1410), by Filippo Maria Visconti (1421–1436), by Doge Pasquale Cicogna (1585–1595), and by Doge Francesco Foscari (1423–1457).

Figure 3.44   Faces of Christ on Serbian silver coins: Stefan Uros I (1243–1276) and Stefan Uros II Milutin (1253–1321).

but with more interpretative freedom and less mintage refinement. The faces on the Serbian coins of Stefan Uros (1243–1321) are quite similar to Venetian *grossi* and depict Christ seated on the throne with long hair, beard, and mustache.

Bulgarian silver coins minted in the fourteenth to fifteenth centuries by Theodor Svetoslav, Ivan Alexander, and Ivan Shishman depict an even more vague Shroud-like face defined by Byzantine canons, and they are even less refined, although they still keep the fundamental features: long hair, beard, and mustache. In some cases, then, some asymmetries typical of the Shroud face are still evident (Fig. 3.45).

Christ depictions endured during the centuries in different states till the present day but never more with the same frequency of representations as during the Byzantine Empire. For example, for celebrating the Great Jubilee in 2000, some states minted coins with the Christ face; among them are noteworthy the 500 *schilling* coins

Figure 3.45   Faces of Christ on Bulgarian silver coins (1300–1400): starting from the left, coins minted by Theodor Svetoslav, Ivan Alexander, and Ivan Shishman.

Figure 3.46 Austrian 500 schilling coin minted for the Great Jubilee in 2000. The face has some weak sindonic-Byzantine reminiscences.

minted by Austria (Fig. 3.46) in which a "modern" Christ is depicted, even though with some sindonic-Byzantine reminiscences.

By carefully observing this face it is clear how the Byzantine tradition, linked to the Shroud, had left a faint sign, especially as far as the beard and mustache are concerned: in this case, the remarkable face asymmetry has almost disappeared, but the beard (torn on the right side) still displays some traces; the hair, though shorter, is still wavy, showing the left curl just in the sindonic position.

The swellings in correspondence to the cheekbones and other sindonic-Byzantine details handed down by tradition during the centuries have instead been lost, but it is interesting to note that after more than a millennium, the face of Christ is still depicted recalling the features of a Byzantine tradition that used the Shroud as a model.

The detailed analysis of the faces carried on up to here allowed us to underline:

- how unlikely it is that an engraver could have depicted a face of Christ on Byzantine coins without having carefully observed the Shroud before, whose image should have been then less deteriorated compared with the image that can be observed today and therefore with several details that were easier to see, like long, asymmetrical hair similar to "payot"

- how this model endured in time and geographical space of several states coming into the Christian tradition, reaching the present day, even though with some attenuations of the somatic details more related to the tortures suffered during the Passion, like tumid cheekbones

The analysis of the faces of Christ above has been carried out from a qualitative point of view, but the following section will confirm the obtained results through a quantitative point of view, with numerical results.

3.5 Quantitative Analysis of the Faces of Christ

To carry out a quantitative analysis of the face of Christ some characteristic parameters have to be taken into consideration. Scientific bibliography [24, 132] suggests several features, some of which are not so easy to identify on the face of the Man of the Shroud.

Among the most evaluated parameters, there are those related to the length of the nose and the distance between the eyes. These will be considered in the present analysis, since the comparison we want to make is that between the peculiar feature of the Shroud face, which appears somewhat elongated with very close eyes, and the different Christ faces depicted on the coins. A dimensional analysis of computerized images has been therefore carried out in order to measure the proportion of the length of the nose compared with the distance between the eyes of the various faces (Fig. 3.47).

From a qualitative point of view, if we observe the variability of the facial features portrayed only on the coins minted by Justinian II during his first period as emperor (Fig. 3.32), we can already note how much the face of Christ is variable and therefore how much this highlights different nose/eyes ratios, without losing some of its fundamental characteristics. Getting on in centuries, this variability increases; therefore the following quantitative analysis becomes of interest.

To understand the meaning of varying this proportion, Fig. 3.48 exemplifies a comparison between the Shroud face, in which the length of the nose is 1.28 times larger than the distance of the

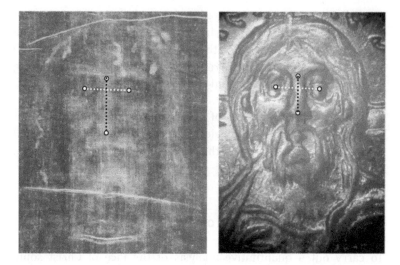

Figure 3.47 Length parameters (distance of the eyes and length of the nose) on the Shroud face (on the left) and on gold money minted by Justinian II (on the right).

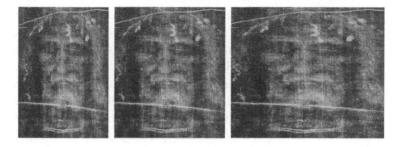

Figure 3.48 Original Shroud face (on the left) with a nose/eyes ratio equal to 1.28, compared with the same distorted face, respectively, in way of obtaining a proportion equal to 1 (in the center) and equal to 0.7 (on the right).

eyes, and two distortions of the same obtained by changing the proportions in 1 and in 0.7, values corresponding to those of some coins. Tables 3.2–3.5 compare the Shroud nose/eyes ratio with the proportion of the nose to the eyes of different faces of Christ on coins minted between the seventh and fifteenth centureis.

Table 3.2 Proportion of the length of the nose compared with the distance of the eyes (nose/eyes ratio) in some Christ faces evaluated by computerized analysis with an approximation of ±0.05, especially due to somatic features (1)

Face picture	Description: emperor & date	Nose/eyes ratio	Notes
	Shroud	1.28	Control image
	Justinian II Period I I mintage 692	1.27	First direct observation
	Justinian II Period I I mintage 692	1.20	First direct observation Different mintage
	Justinian II Period I II mintage 693–695	1.06	Copy from the first mintages
	Justinian II Period I last mintage 695	0.87	Copy from the first mintages

Table 3.3 Proportion of the length of the nose compared with the distance of the eyes (nose/eyes ratio) in some Christ faces evaluated by computerized analysis with an approximation of ±0.05, especially due to somatic features (2)

Face picture	Description: emperor & date	Nose/eyes ratio	Notes
	Michael III 865–867	1.30	Second direct observation
	Basil I and Constantine 867–886	0.86	Copy from other mintages: deteriorated image
	Romanus I and Christophorus 921–931	1.00	Copy from other mintages: deteriorated image
	Constantine VII and Romanus 944	1.20	Third direct observation, Shroud arrival in Constantinople
	Basil II and Constantine VIII 976–1025	1.20	Probable direct observation

Table 3.4 Proportion of the length of the nose compared with the distance of the eyes (nose/eyes ratio) in some Christ faces evaluated by computerized analysis with an approximation of ±0.05, especially due to somatic features (3)

Face picture	Description: emperor & date	Nose/eyes ratio	Notes
	Basil II and Constantine VIII 976–1025	1.13	Beginning of features deterioration
	Romanus III 1028–1034	0.80	Features deterioration
	Michael VII 1071–1078	0.82	Features deterioration
	Manuel I 1143–1180	0.63	Further features deterioration
	Guglielmo I from Palermo 1154–1166	0.71	Features deterioration

Table 3.5 Proportion of the length of the nose compared with the distance of the eyes (nose/eyes ratio) in some Christ faces evaluated by computerized analysis with an approximation of ±0.05, especially due to somatic features (4)

Face picture	Description: emperor & date	Nose/eyes ratio	Notes
	Republic of Dubrovnik (Ragusa)	1.09	Resume of some features
	Lorenzo Tiepolo Republic of Venice 1268–1275	0.85	Features deterioration
	Stefan Uros Serbia 1268–1275	0.87	Features deterioration
	T. Svetoslav Bulgaria 1300–1322	0.60	Further features deterioration
	Manuel II Byzantium 1391–1423	0.80	Features deterioration

It can be noted that, apart from the mintage of Michael III that even exceeds the proportion of the Shroud face, no other coin has the same nose/eyes relation. However, the first coins minted by Justinian II (692) during period I are very close to the Shroud ratio of 1.28 (1.20–1.27), considering the approximation (see below) of ±0.05 due to the measuring method.

This can be explained by the fact that the engraver, conscious of the fact that it was impossible to reproduce the same proportions in a few millimeters, even trying to accentuate the features of the long nose and close eyes, deriving from a careful observation of the Shroud, has perhaps unconsciously preferred being wrong rounding down. On the other hand, the difference between 1.28 and 1.27 is detectable only with the modern means of calculation that were not available at those times.

Already after very few years from the first mintage by Justinian II, the proportion severely decreases to 0.87–1.06, showing that the engravers, not having had the chance to directly observe the Shroud, copied the previous mintages but corrected that ratio that seems not so suitable to a handsome human face.

Instead, something changed with Michael III (around 860), when a Shroud-like face appeared again with an even higher proportion equal to 1.30. This indicates that probably the engravers could see the Shroud (or they more strictly hark back to the canons of the first face of Justinian II).

From that moment on, a deterioration of the typical character-istics of the face of Christ minted on the coins could be observed: nose/eyes ratios reach values even lower than a unit (Basil I: 0.86; Romanus I: 1.00).

Under the reign of Constantine VII, in conjunction with the arrival of the Shroud in Constantinople in 944, the nose/eyes ratio increases again, equal to 1.20, and it remains like that also under the rule of Emperor Basil II (976–1025), although in some coins it decreases down to 1.13.

From 1028, with Emperor Romanus III, the proportion of the nose to the eyes drastically decreases to values varying between 0.63 and 0.82, showing that the engravers depicted the face of Christ with more freedom. Since the Shroud was still in Constantinople, it is possible to hypothesize that in those years, for some reasons, the engravers could not observe the Shroud.

Under the reign of Emperor Manuel I (1143–1180) a face of Christ definitely different from that of the Shroud appeared: a face with short hair and beardless that clearly harks back to that of Christ

Figure 3.49   On the left, the face of a beardless Christ with short hair, minted by Alexius II (1195–1203). On the right, an example of an icon of Emmanuel Christ [2].

Emmanuel (or Manuel), depicting a very young Christ, but with a more aged face in a sign of reverence to the Man-God who is indeed wise like an old man.[8]

It is curious that Emperor Manuel I himself, in addition to the Shroud-like face of Christ, decided to use also the face of the Emmanuel, perhaps in honor of his name. From that moment on, for some decades, the depiction of Christ Emmanuel was alternated with the image of the sindonic Christ; an example is the face displayed in Fig. 3.49, which was minted by Emperor Alexius II (1195–1203).

Of the same period is a Sicilian reproduction (Guglielmo I, 1154–1166) of the face of Christ clearly inspired by Byzantine art with a lower level of details compared to the best coins of Constantinople, but that highlights a nose/eyes ratio equal to 0.71.

Also the Republic of Dubrovnik (Republic of Ragusa) came into contact with the Byzantines, and it is meaningful that between

---

[8]The three little balls, two of them representing the eyes and one on the chin are placed in order to recall the Greek letter delta (Δ) upside down. This feature could have been an iconic reduction of the face of Christ Emmanuel as hypostasis (hierarchical generation of the different realities belonging to the same divine substance) of the Holy Trinity.

1284 and 1372 it issued coins with a Shroud-like face of Christ with a high nose/eyes ratio of 1.09. Even if with a lower level of details, the features of the face seem similar to those of the Shroud. Therefore, the possibility that an engraver of the Maritime Republic of Dubrovnik had the chance of observing the Shroud during travel (perhaps in France) cannot be excluded.

The famous Venetian reproductions of the face of Christ on *grossi* and *sequins*, which lasted for several centuries, were not always very faithful to Byzantine canons. In fact, the faces were more or less similar to the Shroud, but the length of the nose was 0.85 of the eyes' distance. The same can be said for the Serbian *grossi* that depict similar faces to the Venetian coins with analogous nose/eyes ratios.

A further deterioration of the features of the face of Christ can be noted with reference to the Bulgarian coins of the beginnings of the fourteenth century that, besides being less refined in their engraving, reach nose/eyes ratios of even 0.60.

Also in Byzantium, around at the early ages of the fifteenth century with Manuel II, there is a remarkable deterioration in the engraving quality and the Christ face depiction suffers the consequences, though keeping the general features similar to those of the Shroud; the proportion of the nose to the eyes is around 0.80.

To be thorough, in addition to the values of the nose/eyes ratio of Tables 3.2–3.5, the results referring to the two faces of Christ not inspired by the Shroud are also taken in consideration: those on the coins minted by Justinian II in period II, the "Syrian" Christ, with a nose/eyes ratio of 0.90, and the values of the coins by Manuel I–Alexius II, depicting a beardless Christ with 0.58 ratio—very far away from the 1.28 ratio of the sindonic image.

This numismatic quantitative analysis of the face of Christ confirms the previous qualitative observations and perhaps it is more evident in which occasions the engravers could directly observe the Shroud or draw more detailed characteristics on the basis of indirect information.

After this investigation it can be observed that for more than a millennium and a half, till the present day, the face of Jesus Christ has been depicted in a number of coins minted by different empires and states within more than 2000 km (1200 mi) during the early centuries, and all over the world in the present day. Leaving apart some rare exceptions of very different faces of Christ, which probably harked back from other depictions (such as, for example, that portrayed on the coins minted in 1500 in Lucca, Tuscany, that originates from the Holy Face of the city), all these faces have several features in common with one prototype image, that of the Man of the Shroud, minted for the first time by Emperor Justinian II in 692 (Fig. 3.1).

Since, as observed, the face of Christ minted by Justinian II in period I coincides in different details with the face depicted on the Shroud, conclusions can be the following:

(1) The prototype of the face of Christ minted in 692, though being a refined production, does not depict a face with the features of a typical handsome man, because it displays some details linked to the Passion that disfigure the general quality. Why did the engraver carve these features, and why did the Byzantine canon that derived from them codify these peculiar asymmetries and swellings? Obviously the reason is the tortured face of the Man of the Shroud.

(2) The Shroud provided the prototype of the face of Christ that for more than 1000 years has been used in the coinage of several states and empires. This prototype not only developed in the numismatic field but also in painting and sculpture, as observed in Chapter 2 and as demonstrated [40] by other studies.

(3) In conclusion, the Shroud existed in 692 and has been a model for the first coins of Christ, the Byzantine gold *solidi*; this model reached the present day with its general features.

3.6 Critics and Countercritics

As already observed, it is not easy to come to definitive conclusions about Shroud studies, since reasons linked also to one's proper belief lead to infinite pro and con debates, even without having a strong basis for discussions, preventing the possibility to reach clear conclusions.

To show the enormity of the ongoing debates, at the end of this chapter, a discussion about the numismatic theme is reported, and the authors take the chance to critique the interesting conclusions already reached by other researchers.

Some scholars keep sustaining that the Shroud is a work made by a medieval artist and that the face of Christ in the numismatic iconography is independent of the Shroud face. There is no need to prolong discussions with people who deny even the obvious, but as said, later will be considered the less trivial critiques.

The historian Andrea Nicolotti [106] criticized, maybe excessively, *the easiness with which sindonologists sustain the coincidence between the face of the Man of the Shroud and any other Christ portrait with beard and long hair,* referring to *rough operations,* also made by the electronic calculator, *which would be capable to find an analogy between the Shroud face and any other portrait of a vaguely similar bearded man.*

In this case the reference to Whanger's polarized light studies seems clear, though in the text neither the name of the scholar nor the published text are mentioned [152].

Well, it does not seem possible that an international scientific review with qualified editors publishes *rough operations.* On the other hand, it has already been shown that different peculiar characteristics are evidently common to numismatic images and to the Shroud. These features highlight their tight correlation without resorting to computerized methods.

Nicolotti continues in his critique stating that *a fundamental obstacle remains: all these reconstructions start from the premises that the Shroud image is more ancient than all these other images and that these images are modeled on the Shroud—precisely defined "demonstrandum"* (the thesis to be demonstrated).

Maybe the historian forgets that to this day the Shroud body image is not technically reproducible and not scientifically explainable yet. This is an important point because it underlines how difficult it is, rather impossible, to make such a reproduction. Supposing that in the past, let us say in the Middle Ages, there was an artist able to reproduce a similar image enveloping a tortured corpse. How could the artist depict also all the iconographic details (swollen cheekbones and eyebrows, torn beard, etc.) identified in this numismatic study? With what cynicism should he have pre-emptively prepared the corpse?

It has also to be added that, obviously, in the Middle Ages all the numismatic depictions of Christ today available were not known; therefore the hypothetical artist had no way to copy something that has been discovered centuries afterward. The peculiar features of the Christ faces depicted on the Byzantine coins could not be known by a hypothetical medieval artist because, even in 1845, the faces portrayed on the first coins minted by Justinian II were not known yet. In fact, the numismatist Giulio di San Quintino [41], in one of his works, makes a list of the several coins of Emperor Justinian II, highlighting the presence of gold coins depicting the face of Christ of type II of period I and of period II in the museums of Milan and Paris, but he does not name any coin of type I of period I; so all the particulars highlighted in Fig. 3.34 were not even known two centuries ago, least of all in the Middle Ages.

In the light of these considerations, the "demonstrandum" (the thesis to be demonstrated) automatically becomes "demonstratum" (the demonstration of the thesis).

Finally, Nicolotti affirms that "...there was not neither a strictly codified iconography, nor a unique model ..." as far as the face of Christ on Byzantine coins is concerned, and to corroborate this statement he recalls the "so-called Syrian face of Christ" minted by Justinian II during his second period of ruling. Nicolotti perhaps forgets the beardless face of Christ minted by Manuel I Comnenus, which is not mentioned in his book. The presence of this other face confirms the fact that Byzantines did not feel themselves obliged to follow only the sindonic Christ model. Then, it has to be taken into consideration also that at that age there were other images of Christ, like the Image of Camuliana, and it is likely that some emperors would have preferred to refer to those ones, maybe for celebrating some anniversaries or for coincidence of names (Manuel I was the first one to reproduce Christ Emmanuel).

The statement referring to the absence of a strictly codified iconography seems, instead, not reliable, both because the eastern icons actually demonstrate the existence of a rigorous canon to which to make reference for every depiction and because, as shown, several numismatic details recur in a number of Christ faces and last during the centuries till the present day.

On the contrary, it is interesting to observe that the young and beardless Christ iconography, developed during the early centuries in Rome, suddenly disappeared around the fourth and fifth centuries when the physical features of the sindonic Christ spread throughout the Roman Empire.

Another scholar against the conclusions reached by the authors is the art historian Davor Aslanovski [11, 12] from Oxford University, who is involved in the issue of the face of Christ in ancient history and kindly accepted to have a telematic debate with G. Fanti. Aslanovski states that "nowadays there's no serious historian who thinks that Byzantine icons show the so-called Vignon markings." Actually, some marks are quite subjective, like the U shape between the eyebrows or the V triangle on the nose bridge, but it is not appropriate to paint every element with the same brush. As already said, it is convenient to start from the critical analysis of the Vignon markings in order to define some more objective characteristics that have been used in the numismatic comparison.

Aslanovski observes that if two things resemble each other, it does not necessarily mean that they are linked together. This observation can be acceptable for some mintages that depict the face of Christ approximately (like, for example, the Venetian, Serbian, and Bulgarian *grossi matapan* that reproduce Byzantine canons more vaguely), but it becomes more difficult to make this kind of statements with reference to the mintage of Justinian II, type I, period I, in Fig. 3.33, in which there are too many "coincidences" to think something like that. There are, then, some objective characteristics, such as the hair asymmetry like "payot" on the shoulders (at the moment no other analogous hair asymmetries have been found in other representations of Byzantine emperors) or the skewed nose, that cannot be called into question.

The historian goes on affirming that, even if everybody agrees on the resemblance, this does not prove anything, except that there is a resemblance. Although the comparison itself cannot prove more than this, if we combine all the several resemblances (and the mentioned scholar Whanger found more than a hundred of them in some mintages), the result leads to the more obvious conclusion that some Byzantine engravers should have directly observed the Shroud.

Aslanovski adds that, if the resemblances are used as evidence, it is necessary to check if there are also some differences. This is completely correct; however, when the first author asked him to provide a list of the different aspects, he did not receive any reply, except for a reference to the hypothetical differences between the nose/eyes ratios (which for the Shroud he declared equal to 1.4 instead of 1.28) and to the angle of the eyebrow with reference to a work published in 1987 [43].

If we observe the variability of the facial features displayed just on the coins minted by Justinian II during his first period (Fig. 3.32), it can be seen how easy is to find a face of Christ with different ratios compared to the Shroud face without entailing the loss of some fundamental features.

Furthermore, the measurements considered as different by Aslanovski (face of the Shroud with a declared ratio equal to 1.40) have been evaluated with reference to segments of only 3 mm, probably not by means of a computerized analysis like in the Tables 3.2–3.5, but using more rough tools like a double decimeter, causing high uncertainty.[9] Finally, these measurements depend on the reference chosen for defining the segments and the subjective choice, for example, of the upper limit of the nose, can cause high doubts.

The nose length and eye distance ratio of the Man of the Shroud is 1.28 ($\pm$0.05) and it is compatible[10] (Fig. 3.50) with that of the Byzantine coins of Justinian II of the first period, equal to 1.27 ($\pm$0.05), as well as with the successive coins of Michael III equal to 1.30 ($\pm$0.05). As far as the angle of the eyebrow is concerned, the uncertainty related to these measurements is so high as to prevent any possibility of carrying out reliable comparisons; on the other hand, relevant differences cannot be detected by the naked eye.

---

[9]This uncertainty will be discussed in Chapter 4; the reader can refer to that chapter.
[10]Given more values measured with respective bands of possible variability (uncertainty) their compatibility, in metrology, is a condition that occurs when at least one element belonging to the correspondent band of values is common to all the other bands.

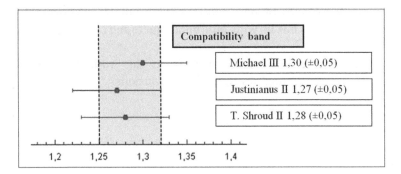

Figure 3.50 Compatibility band of the Shroud face nose/eyes ratio and of two Byzantine *solidi* by Justinian II and Michael III: all the values included between 1.27 and 1.30 are acceptable for the three portraits.

Referring to the asymmetrical "payot" Aslanovski proposes the hypothesis that they derive from "artistic imagination." How can be thought to this peculiar "imagination" concerning not only an artist but tens or even hundreds of artists who for several centuries depicted these features? And actually how can we do this in a Byzantine iconographical environment where everything was ruled according to specific canons?

Aslanovski objects that the asymmetrical "payot" are a very bad subject, because in copying a face, nobody pays attention to the hair. On the contrary, this observation strengthens even more the correspondence between the Shroud face and the Byzantine faces of Christ because, as it can be observed by the comparison in Fig. 3.33, not only the "payot" but also the more evident waves are depicted. This demonstrates how much attention the engraver paid in the observation of the face of the Man of the Shroud.

Aslanovski concludes by stating that for establishing a link, something more than a resemblance is necessary, and *it is for this reason that I defined the sindonologists' interpretations—pure fantasy—because there are no evidences.*

It is clear that nobody can talk of a 100% sure evidence, but from a statistical evaluation on the whole set of "coincidences" reported in Section A.3 of the appendix we can reach a certainty equal to 99.9999999999999993%. Not just for this minimum residual of 0.0000000000000007% it seems possible to accuse the sindonologists of pure fantasy, but when the iconographic-historical issues mix up with faith, minds ignite and can reach absurd conclusions.

Anyway, from the data reported in this wide numismatic overview anyone can draw their own conclusions.

Part II

The Fascinating Dating Quest

Chapter 4

Radiocarbon "Distraction"

In the eighties, also thanks to the important scientific research carried out on the Shroud by the Shroud of Turin Research Project (STURP) , a group of about 40 American scientists), a lot of evidence in favor of the authenticity of the relic had been published, insomuch that there was the almost certain proof that the sheet enveloped the corpse of Jesus.

In addition to the well-known fact that the body image is visible to the photograph negative, its three-dimensionality was proven, and there had been the confirmation that the image could not be reproduced with the most sophisticated modern techniques (these issues have been developed in Chapter 1). Besides, the Swiss palynologist[1] Max Frei retraced the Shroud historical trail starting from Jerusalem on the basis of the pollen taken from the relic and someone even declared of having discovered the imprint of a *dilepton lituus*, a coin minted around A.D. 30 under Pontius Pilate, on the right eye of the Shroud.

After having verified also the perfect correspondence between the signs of the Passion of Jesus Christ described in the Gospels and

---

[1]Palynology is the science that studies pollen and other microscopic biological elements (spores of mosses, club and ferns, fungal spores and parts, etc.), both current and fossil.

The Shroud of Turin: First Century after Christ!
Giulio Fanti and Pierandrea Malfi
Copyright © 2015 Pan Stanford Publishing Pte. Ltd.
ISBN 978-981-4669-12-2 (Hardcover), 978-981-4669-13-9 (eBook)
www.panstanford.com

the tortures suffered by the Man of the Shroud, only one single test was still missed in order to confirm the hypothesis that the Shroud actually wrapped the body of the Resurrected, that is, the evidence concerning the date of the cloth, through a method that had been developed precisely in those years: radiocarbon 14 dating. However, the outcome diverted the scientists from the way of scientifically proving what the Gospels ask to accept by faith.

After a brief mention regarding the radiocarbon dating method, this chapter will describe how this test was carried out and what results have been obtained in the light of the various critics arisen. With relation to this last issue, it has to be observed that the burning debate among scientists in favor and against the validity of the data obtained by the means of 1988 radiocarbon dating rests to a large extent on particularly technical aspects of statistical analysis and metrology.[2]

4.1 The Carbon 14 Method: An Overview

The atomic nuclei have an interesting peculiarity: they are not all equal, and even the nuclei of the same element can be different each other. Let us take, for example, carbon, hydrogen, and helium. Each hydrogen atom has in its nucleus one single proton, each atom of helium has two protons, and any carbon atom has six protons. In other words, the number of protons in the nucleus of an atom univocally defines the element: if there are 20 protons, we have a calcium atom; if protons are 92, we have an atom of uranium, etc.

In the nucleus there is also a particle with no electric charge, the neutron. It is quite surprising that, once the number of protons in the nucleus is fixed, that is, once the chemical element is fixed, the number of neutrons in its nuclei is not always constant but slightly varies from a maximum to a minimum value. For example,

---

[2]From the International Vocabulary of Metrology: Metrology is the science of measurement and its application. It includes all theoretical and practical aspects of measurement, whatever the measurement uncertainty and field of application. Measurement is the process of experimentally obtaining one or more quantity values that can reasonably be attributed to a quantity. Quantity is the property of a phenomenon, body, or substance, where the property has a magnitude that can be expressed as a number and a reference.

the chlorine element (17 protons) has both stable atoms with 18 neutrons and atoms with 20 neutrons. In technical language it is said that chlorine has two stable isotopes.[3]

Instead, carbon, which is contained in all organic substances, is constituted of 99.89% of carbon 12 (^{12}C) and of 1.11% of carbon 13 (^{13}C); carbon 14, ^{14}C, that is, the radioactive isotope of carbon that is involved in the dating process, is in a much more reduced percentage: there is about one atom of ^{14}C in every trillion atoms of ^{12}C (more precisely[4] 1 out of $8 \cdot 10^{11}$ atoms of ^{12}C); even more rare isotopes (^{10}C, ^{11}C) exist, unstable too, but in a negligible percentage.

Radiocarbon 14 dating is based on the calculation of the percentage of these radioactive atoms compared to the number of atoms of ^{12}C in the sample analyzed. The sample is then dated on the basis of a relationship that links the measured percentage to an estimated date.

The radiocarbon dating method was invented by Willard F. Libby [93] between 1946 and 1955, and it is based on the following facts. ^{14}C is being produced in the upper Earth's atmosphere through the collision with the cosmic radiation, which generates high-kinetic-energy neutrons, which, in turn, strike atmospheric nitrogen atoms;

---

[3]The word "isotope," from Greek, literally means "the same place." The number of protons in the atoms being always the same, independently from the number of neutrons in the nucleus, different isotopes of a single element occupy the same position on the periodic table, from which came the name "isotope," because protons, not neutrons, are those that in the nucleus distinguish an element from another. In science, a particular notation is used to indicate an isotope. For example, ^{35}Cl, ^{37}Cl (to be read as Chlorine 35 and Chlorine 37); the number in the superscript at the upper left of the chemical symbol is equal to the sum of protons and neutrons in an atomic nucleus. Since chlorine has 17 protons in its nucleus, by subtraction it results that ^{35}Cl has 18 neutrons in the nucleus, whereas ^{37}Cl has 20 neutrons. Chemical elements are to a large extent constituted by a mixture of stable isotopes in which each of them has a certain percent abundance, that is, how many atoms in 100 have a kind of isotope, how many have another, etc. For example, the percentage abundance of ^{35}Cl is 75.77%, while that of ^{37}Cl is 24.23%.

[4]This amount can seem too small to be significant; however, it is useful to remind the readers that any living being is constituted of an extremely high number of carbon atoms. Let us consider this quick exemplifying calculation: in 12 g (0.42 oz.) of carbon there are about 600,000 billion of billions of ^{12}C (equal to $6.022 \cdot 10^{23}$). Since there is about one atom of ^{14}C every trillion of atoms of ^{12}C ($1.0 \cdot 10^{12}$), then, in the above mentioned 12 g there are more or less 600 billion of ^{14}C that are not impossible to be experimentally detected.

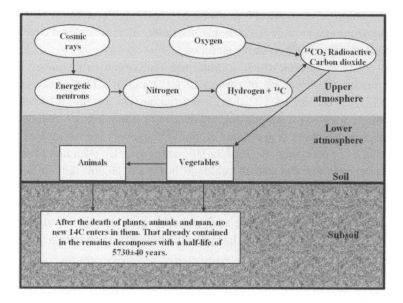

Figure 4.1 The carbon 14 cycle. Particular interactions in the upper atmosphere between high-energy cosmic radiation and nitrogen atoms lead to the formation of radioactive carbon dioxide ($^{14}CO_2$) as in this element there is the ^{12}C isotope instead of the ^{12}C. This radioactive $^{14}CO_2$ joins the life cycle of plants and animals, including humans. The method, exploitable on artifacts deriving from living beings, allows one to date back the object on the basis of the count of $^{14}CO_2$ atoms.

they change into hydrogen and ^{14}C, the latter combining with oxygen atoms to form radioactive carbon dioxide ($^{14}CO_2$). Like the normal carbon dioxide, it is incorporated into plants by photosynthesis and acquired by animals that eat vegetables. Carnivores assimilate $^{14}CO_2$ as well because they also feed on herbivores (Fig. 4.1).

The disintegration rate of ^{14}C is almost constant, and its quantity is reduced by half in 5730±40 years.[5] The symbol ± associated with

---

[5]The quantity of ^{14}C decreases according to the law of radioactive decay $N(t) = N_0 e^{-\lambda t}$, where N is the number of radioactive isotopes; "half-life" $t_{\frac{1}{2}}$ (equal to 5730 years for ^{14}C) is given by $t_{\frac{1}{2}} = \frac{ln2}{\lambda}$ and $\lambda = 0.00012097$. This period is known with the technical term "half-life," that is, the time taken for ^{14}C to decay to half of its initial value and to transform the half of the initial amount of ^{14}C contained in the sample in ^{12}C.

the time of 40 years corresponds to the uncertainty assigned to the measured value of 5730 years.

As already anticipated, in this chapter, as well as in those (Chapters 6 and 7) dedicated to alternative dating methods, it is necessary to use some fundamental concepts of metrology, essential to understanding the debate. The first concept is uncertainty, which is the meaning of the number that follows the symbol $\pm$ after a measured or calculated value.

The basic idea of uncertainty is not new and is often used at a qualitative level when, for example, it is said that a room has a surface of *about* 430 square feet or that someone weights *around* 150 lb. The message conveyed is that the data provided is not effectively exact and a certain deviation is admitted. From the scientific point of view, however, a deviation has to be objectively quantified ("about" and "around" are subjective expressions). So, all the uncertainty is giving a quantitative and objective sense to the deviation associated with a parameter.

Uncertainty is technically defined as "the parameter, associated with the result of a measurement, that characterizes the dispersion of the values that could reasonably be attributed to the measurand." Uncertainty provides therefore an interval of values that may be expected to encompass the measured value. The validity of the estimate is declared by referring to statistical calculations and by indicating the so-called confidence level; usually uncertainty is given with a 95% level of confidence.

It is likely that readers with no familiarity with metrology could find some difficulty in understanding the paragraph above. The following example will try to clarify it.

Let us measure for 100 times the width of a long ancient table that can be considered to have a rectangular board, even if the signs of time made its long sides not exactly parallel each other. After having taken the width measurements from different positions, we need to enter the resulting data in a calculator capable of statistically calculating the uncertainty. Let us suppose that all the values measured are between 95 cm (3.117 ft.) and 105 cm (3.445 ft.) with a more frequent number of data equal to 100 cm (3.281 ft.). The calculator provides, for example, an average width value of 100 cm and a deviation (called standard deviation) equal to 2 cm (0.066 ft.).

Further, let us assume that from the statistical point of view data could follow a particular distribution called Gaussian distribution.[6] At this stage, the rules for calculating uncertainty, which we do not mention, lead to the affirmation that the table width is 100 ± 4 cm (95%). In the light of the example, let us read again the technical definition of uncertainty provided above.

- *Uncertainty provides an interval of values that may be expected to encompass the measured value.* In fact the uncertainty of ± 4 cm defines an interval: the table width is between 96 and 104 cm.
- *The validity of the estimate is declared by referring to statistical calculations and by indicating the confidence level in reference to a Gaussian distribution; usually uncertainty is given with a 95% level of confidence.*
  In the example, the calculator provided a deviation of 2 cm, whereas the uncertainty has been calculated as equal to 4 cm, which is double the amount of deviation. This is because the 95% level of confidence has been assumed in the context of the Gaussian distribution; if the statistical distribution associated with data changes, the calculation modalities will be different. Anyway, that 95% in brackets means that only 5 measurements out of the 100 performed exceed by defect the value of 96 cm or by excess the value of 104 cm, or, rather, that 95% of data is in the interval 100 ± 4 cm.

As far as the ^{14}C half-life is concerned, 5730 ± 40 years (95%), it is encompassed between 5690 and 5770 years, with a "validity of the estimate" of 95%; this means that the half-life of 95% of ^{14}C atoms is situated in that interval. Having this information, how can an artifact deriving from living beings be dated?

Each living being constantly absorbs during time a determined amount of ^{14}C. Until a plant, an animal, or a human is alive, the ^{14}C that decays in the body is substituted by other ^{14}C; so there is a balance between the ^{14}C decomposing and the ^{14}C that has

---

[6]This means that data tend to concentrate around the average value, 100 cm in this example, so that in 100 width measurements of the table, values of 101 cm and 99 cm are more frequent than values of 102 or 98 and values of 104 or 96 are much less frequent.

been assimilated. Once they die, they cease to acquire ^{14}C, and the ratio of ^{14}C to ^{12}C in the remains will gradually reduce. The method allows, by determining the ^{14}C residual percentage compared to ^{12}C, to calculate how long it has been since a given sample stopped exchanging carbon, under the condition that the artifact has always been conserved in an environment without external contaminations. Being aware, in fact, that the half-life of ^{14}C quantity is 5730±40 years it is possible to assign a date correspondent to the death of the living being.

The existence of the radiocarbon dating technique is rather known, if for no other reason than it is often cited in the explanation cards of some artifacts exposed in museums; sometimes writing says "artifact dated using radiocarbon method." In the collective imagination this test is the artifacts' dating method par excellence, well proven and reliable, essentially shrouded in an aura of infallibility.

However, like all the other scientific methods of analysis, it is based on some "validity rules," which means that some hypotheses have to be verified in order that the obtained results can be considered as valid. Behind a scientific method there are various models, which are schematizations that, as far as complex they can be, they have some limits in describing real, often very complicated, behaviors.

Think, for example, of weather forecasts: the more they move forward in time, the less the forecast is reliable, because there are too many variants that can change or overstep the behavior limits assumed in the reference model from which data are elaborated. Ultimately, out of the rules, a method of analysis, whatever it is, loses its validity.

Here is, to a large extent, the heart of the matter of the Shroud radiocarbon dating: are all the validity rules of the method respected? Not really, as it will be shown below.

What are these rules then? The ^{14}C method is based on the following fundamental hypotheses [51]:

(1) The ^{14}C percentage of the atmosphere at the moment of the death of the sample is known.

(2) All the organs or tissues of the sample absorb the same percentage of ^{14}C.

(3) There are no environmental contaminations of the sample under analysis.[7]

(4) The sample taken for the analyses has to be representative of the whole vegetable or animal under investigation.[8]

4.2 Application of the Radiocarbon Dating Method to the Shroud

The radiocarbon dating of the Shroud was and is, for the time being, not recommended. In fact the different places in which it has been exposed and of which there is no historical trace, the different contaminations during time, also during public exhibitions, the fires, and perhaps also the radiations[9] it was subject to during the body image formation, do not satisfy the fundamental assumptions of the method.

In particular, the third hypothesis requires that the artifact be not contaminated, but even if today the radiocarbon dating were repeated, it would not be clear which kind of pre-emptive cleaning applies: this is because both the environmental factors and the body image formation mechanism that could have interact with the sheet are not known.

The week before that the 1988 radiocarbon dating result was distributed by the review in *Nature*, the Harvard physicist T. J. Phillips published an article [112] in which he highlighted that if the Shroud were the burial cloth of Jesus of Nazareth, it could have been actually irradiated with neutrons that could have modified the ^{14}C percentage.

If a trustworthy radiocarbon dating of the Shroud has to be carried out with the ^{14}C method, it is evident that, first of all,

---

[7] For example, when Willy Wölfly was the director of the laboratory at the Zurich Polytechnic Institute, a tablecloth bought 50 years before was dated with the radiocarbon method; the tablecloth was found to be 350 years old.

[8] For example, if a big arras kept enrolled for centuries had to be dated, it would not be correct to take a sample from the external side, since it could have been contaminated by the surrounding environment.

[9] According to the most reliable hypothesis, the body image formation was caused by a not yet defined "radiation" coming from the inner of the enveloped body, which could have interfered on the ^{14}C percentage.

science has to try to explain the body image formation process; only after having verified the hypothesis with documented experimental procedures, it will be possible to think about the way in which performing a radiocarbon 14 dating test that could provide reliable results.

It is important not to forget, finally, that during the sample taking operations in 1988 (accomplished without gloves, so with an obvious external contamination), to reduce the contamination by acari and bacteria, it has been decided to put the empty reliquary in a plastic bag containing several grams of thymol [101] for some hours. According to some scholars, among which was R. Rogers,[10] thymol does not completely evaporate; once the reliquary was closed again with the Shroud, this substance had contaminated, perhaps seriously, the linen cellulose. Thymol, in fact, reacts with linen cellulose and yellows it.[11] Even worse, the phenol (C_6H_5OH) of the thymol is absorbed by the cellulose that is enriched in this way of "young" carbon: a potential future dating, almost certainly, will provide a more recent date because of this contamination.[12]

Though some reasonable objections had been shown, on April 21, 1988, a sample was taken by the corner of the frontal imprint of the Shroud cloth, on the side of the left foot (Fig. 4.2) [73].

The strip had dimensions of 81 × 21 mm (3.19 × 0.83 in.) and weighed 447.5 mg (0.01579 oz.). Its edges had been removed and then it was cut in halves; one of the two segments, with dimensions of 38 × 21 mm (1.50 × 0.83 in.) was then divided into three parts (Fig. 4.3).

At a later stage, each lab divided the samples into a given number of subsamples and carried out a series of different dating tests, used for a laboratory statistical analysis, coordinated by Professor M. Tite from the London British Museum.

---

[10] Rogers R., private conversation, 2003.

[11] Thymol, causing the yellowing of the cellulose, reduces the existing contrast between body image and background, making even less visible the already faded imprinted image.

[12] At this point a question is necessary: how many times in the past had the Shroud been disinfested from microorganisms or "better" conserved by exposing it to substances similar to thymol that changed the isotopic percentage of ^{14}C?

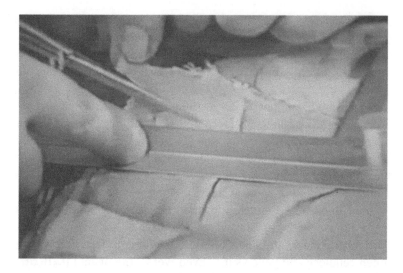

Figure 4.2 Sample taking (without gloves) by the Shroud cloth for radiocarbon dating (G. Riggi di Numana, Fototeca 3M).

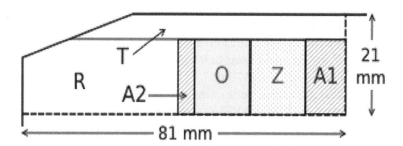

Figure 4.3 Model of the strip of the Shroud used for the radiocarbon test with the symbols of the samples given to the different labs; O = Oxford; Z = Zurich; A1+A2 = Arizona; R = "Riserva"; T = cut segment.

The outcome had been announced officially, not by the scientists who carried out the analyses, but by the director of the Vatican Press Office, M. D. Joaquin Navarro Valls, on the basis of a press release issued by Cardinal A. Ballestrero and corrected by the Holy See, on October 12, 1988 [73, p. 168]. The public announcement stated that according to the documents provided by the laboratories:

The calibrated calendar age range assigned to the Shroud cloth with 95% confidence level is from 1260 to 1390 AD . . . In submitting to science the evaluation of these results, the Church confirms its respect and veneration for this venerable icon of Christ, which remains an object of devotion for the faithful in keeping with the attitude always expressed in regard to the Holy Shroud, namely the value of the image is more important than the date of the Shroud itself . . . At the same time, the problems about the origin of the image and its preservation still remain to a large extent unsolved and will require further research and study. In regard to this, the Church will show the same openness, inspired by the love of truth which it showed by permitting the radiocarbon dating as soon as it was presented with a reasonable and effective program in regard to that matter.

Unfortunately, the previously proposed protocol was not respected, and this gave rise to all the possible conjectures against a scientifically correct procedure.

4.3 A Result Nonresult

The absolutely unexpected test results left everybody astonished because they were contradictory with regard to the amount of scientific evidence in favor of the authenticity discovered a few years before. The out-of-the-blue assignation of a medieval date to the Shroud upset both the scientific community and the world public opinion. Some scholars, such as Eric Jumper, belonging to the famous STURP and blindly trusting in science and in the radiocarbon dating method, quit their studies on the Turin relic.

The result was an unexpected victory for all the people who, for different reasons, were irritated by the idea that the Shroud could be original; people convinced of the relic's authenticity, also on the basis of the recent discovered evidence, but who blindly trusted in science and in the radiocarbon dating method, experienced a great crisis. However, not everyone passively accepted this outcome by putting into the foreground the radiocarbon method in respect to the amount of opposite evidence.

Controversy grew acrimonious, and all the possible, more or less scientific, explanations were developed in order to justify the medieval result. Still in the present day, there is someone who even accuse the scientists of having substituted the samples for obtaining the predetermined result.

Someone observed that the Shroud is not the only fabric that could be contaminated by external environmental effects. For example, the radiocarbon dating of Egyptian mummy n. 1770 kept in the Manchester Museum [45, pp. 119–131] provided different ages for the bones and the bandages; these last, covered with a biologic coat, according to the scholar Leoncio Garza-Valdes,[13] were 800–1000 years "younger" than the bones.[14] If the radiocarbon dates were trustworthy, it would lead to the absurd conclusion that the corpse would have been enveloped in the bandages 800–1000 years after his death!

Very soon, the statistical calculations of the results related to radiocarbon test published by the review in *Nature* [44] were double-checked, and serious mistakes had been found. For example, engineer Ernesto Brunati, [29, pp. 51–52], [30, p. 37], [31], observed that the statistical parameter of the *significance level*,[15] published in reference to the Shroud dating, is not 5% but 4.17%; therefore results had not to be combined each other but carefully re-examined.[16] Furthermore, making a new count of the average ages

---

[13] L. Garza-Valdes is a scholar at University of Texas Health Science Center of San Antonio, USA.

[14] Also for this case, like for many others, the dating is debated; for example, someone reports a difference between the date of the linen and the date of the bones of only 340 years (*Radiocarbon*, Vol. 24, No. 3, 1982, pp. 262–290) and supposes that the mummy had been wrapped again some centuries afterward.

[15] Significance level corresponds to:

$$s = e^{-\frac{\chi^2}{2}}$$

where χ^2 is calculated according to the Ward and Wilson formula:

$$\chi^2 = \sum_{i=1}^{n} \frac{(x_i - m)^2}{\epsilon_i^2}$$

in which m is the arithmetic mean of the sample and ε is the uncertainty of the sample connected to the repeatability.

[16] Significance level s indicates if the variability of the measured values depends only on the repeatability or if it is likely instead that also bias components are present; s varies between 0 and 1. If s tends to 0, from the statistical point of view it can be stated that the variability of the measured data is not imputable only to random

on the basis of the data published in *Nature*, Table 1, he obtained that the value of the sample of Arizona was different from that published, reducing the significance level to 1,04%. To put it simply, this significance level indicates that there are about 99 odds out of 100 that the radiocarbon result is not reliable.

According to the scholar Remi Van Haelst [143], who reanalyzed the statistical calculations on the basis of the data published in *Nature* [44], the correct conclusion, which should replace the existing one, should be:

The results of radiocarbon measurements of Arizona, Oxford and Zurich yield a calibrated date of 1280–1300 with only a significance level of 1.2%. These results therefore furnish the conclusive evidence that the samples used by labs are NOT homogeneous in C-14 content.

According to Leoncio Garza-Valdes [70], there would be a bioplastic coating that covered the Shroud linen fibers, which would make difficult any cleaning method of the sample. So, in this case, the third fundamental hypothesis that states that the sample has to be free from any contaminations is not verified. He affirmed to have identified, on a Shroud sample, the existence of a biologic complex composed of fungi and bacteria, which covers as a coating the linen fibers and cannot be removed by the conventional cleaning methods of most carbon dating labs; this therefore would have altered the radiocarbon dating. This so-called bioplastic coating could be that visible in Fig. 4.4, corresponding to a Shroud fiber observed by an optical microscope.

The American researcher R. Villarreal [147], analyzing a thread declared coming from the Shroud, provided by R. Rogers and extracted by Professor Luigi Gonella Gonella from the sample taking in 1988, discovered that one end was made of linen but, instead, the other end was made of cotton. This would confirm the hypothesis, sustained by some scholars, that the majority of the samples taken from the Shroud for radiocarbon dating derived from a so-called invisible medieval patching. The fragment of the linen

---

effects, but also to a not negligible bias. Generally the limit of significance level equal to 0.05 is used as a convention. If, like in the case under examination, a significance level obtained is lower to this limit of 0.05, that is, equal to 0.0417 (or 4.17%), the value *s* is significant and the existence of a not negligible bias (environmental, in the radiocarbon case) has to be hypothesized.

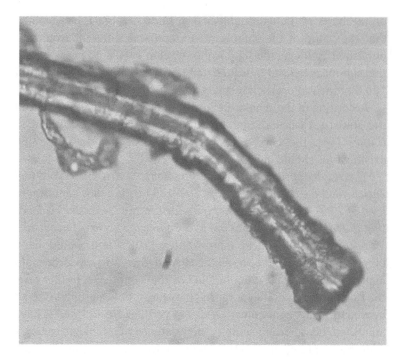

Figure 4.4   Shroud fiber covered with contaminating material.

thread analyzed would belong to the Shroud, whereas the other end would be the result of the hypothesized patching. However, someone doubted that the thread taken by Prof. Gonella derived from a top-stitched hem of a not-well-determined period.

There is also another hypothesis that states that the whole sample taken from the 1988 test was part of a repair made in the thirteenth century and not after the 1532 fire. This hypothesis would therefore explain why the Chambèry fire holes had not been repaired. Did there perhaps exist a document at the time of the sampling that reported this repair and that was used by someone to select the 1988 sampling area?

These and other "explanations" of the radiocarbon dating outcomes were not enough for convincing a great number of scientists, who considered the result as reliable and stated that, even if there are some unclear points, the result clearly showed that the Shroud could not have been sewn in the first century

A.D. Among these, for example, there was the scientific advisor of the Guardian of the Shroud, Prof. Gonella, who admitted[17] that the statistical results published in *Nature* were not correct and that the uncertainty assigned to the result should have been increased, but despite this, the medieval result could not be put into question.

It is obvious that if there is an error in a scientific evidence, this will have to be radically removed by doing the test completely again, and not approximately only by increasing the uncertainty assigned to the result. In the previous case it is true that results lead to dates between 1155 and 1359, therefore medieval dates, but the placement of these values on a few-centimeter piece of fabric makes one think of a trend, maybe due to external environmental factors, that could vary the date of a cloth more than 4 m in length, even by millenniums. It is necessary then to recognize and remove the *bias* first and then repeat the test with the right scientific criteria.

The concept of bias (systematic measurement error) is important not just in relation with the radiocarbon dating test but also in the context of the alternative methods of dating introduced in Chapters 6 and 7. Bias is a variation of the property analyzed in the sample under examination that does not derive from random phenomena but from an external factor able to modify the expected value: for example, the misalignment of the rifle gun sight with respect to the barrel systematically produces a wrong trajectory. If it is known, the bias can be corrected and therefore canceled; in fact, it is sufficient to aim to a distance from the center that is equal and opposite the gun sight misalignment.

Another example: All the body weight measurements taken by the members of a family are affected by the systematic effect if the scale has not been reset (i.e., if its needle is not placed on 0 when it is unloaded). Once that is realized, it is not necessary to take again all the measurements; it is simply enough to correct the data, deducting the bias. So, if unloaded the scale points at 2 kg, then a measurement of 84 kg has to be corrected to 82 kg, that is, 84 less the bias (2 kg).

Several years later, the Shroud science[18] scholar Gerardo Ballabio proposed a spatial distribution of the dates resulting from each

---

[17]Conversation with the first author.
[18]For the description of this group of scholars the reader can refer to the end of Chapter 2.

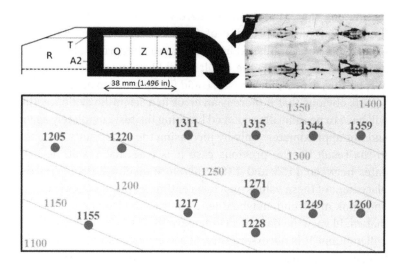

Figure 4.5 Possible spatial configuration of the dates (dark circle) of the subsamples deriving from the piece of the Shroud taken for the radiocarbon test and dated by the three labs: Oxford (O), Zurich (Z), and Tucson in Arizona (A1 and A2). The fragment R on the left is the so-called Reserve—the part retained in case further material was necessary. The part indicated with T is the nonsignificant edge of the fabric. The dashed lines indicate the subsamples' placement. A contamination is evident that in 6 sq. cm (0.093 sq. in.) of fabric causes a bias of 204 years.

single sample of the Shroud displayed in Fig. 4.5, in which a trend of the date can be noted that depends on the physical position of each fabric sample.

First of all, this result shows, without any elaborate statistical evaluation needed, that the uncertainty assigned by the *Nature* paper, at a 95% confidence level, equal to ±65 years, is not correct, because in 12 samples there is a variation of 204 years, which is far more than the 130 years considered in the report.

Beyond this, the variability of the resulting dates shows that the few centimeters of fabric, taken from the Shroud, cannot be representative of the whole, longer than 4 m (13 ft.), cloth. In fact, the fourth fundamental hypothesis states that the sample taken has to be representative of the entire fabric. If in just a few centimeters a variation of more than two centuries is observed, it is clear that the hypothetical dating of each piece of such a big cloth would furnish far more different dates, and so the few-centimeter piece taken from a corner of the cloth is not representative of the whole Shroud. To do something better it would be necessary to date samples coming from different areas of the sheet, but obviously it is not possible to riddle the Shroud with holes. Here is one reason that explains why a second radiocarbon dating test of the relic has not been performed yet.

Starting from these observations, a group of professors of statistics at Parma (M. Riani), London (A. Atkinsons), and Udine (F. Crosilla), in cooperation with the first author, demonstrated that the 1988 radiocarbon dating result referring to the Shroud is not reliable.

These results have been first published in the Italian Society of Statistics review [121], in which, among others, conclusions are:

> *The statements of Damon et al. (1988) that <<the quoted errors reflect all sources of error>> and that <<the results provide conclusive evidence that the linen of the Shroud of Turin is mediaeval>> need to be reconsidered in the light of the evidence produced by our use of robust statistical technique. In other words, the 12 measurements produced by the 3 labs cannot be considered as repeated measurements of a single unknown quantity, therefore an environmental contamination in the analyzed piece of fabric that acted in a non-uniform linear way, adding a non-negligible bias, can be hypothesized.*

Conclusions are very important:

> *If the bias highlighted by the radiocarbon dating of the three labs was directly transferred all along the Shroud, it could be hypothesized, for a length of about 4 meters, a variation of twenty millenniums in the future, starting from a date of the edge dating back to the first millennium A.D.*

This conclusion affirms therefore that the result of the 1988 radiocarbon dating is unreliable and scientifically meaningless.

Analogous results have been later published by the same authors in a prestigious international statistics review [120]: After a "robust statistical analysis" of 387,072 different possible configurations of the distribution of the 12 subsamples of the strip taken from the Shroud, conclusions are that the data relative to the carbon dating "show surprising heterogeneity" and the resulting trend could be due to an environmental bias.

The bias producing this date variation is not defined yet, but it seems obvious that if this bias is not uniform along the Sheet, dates are heterogeneous. For example, let us think about pouring some oil drops on a big sheet folded in several layers, and let us suppose that the oil could vary the radiocarbon percentage of the threads: the oil will stain the paper surface heterogeneously. If some samples are then taken in order to make a radiocarbon dating test, it is easy to think how much the resulting dates will be heterogeneous.

In agreement with the previous work, and in support of the reliability of robust statistical analysis carried out on 12 subsamples, it turned out that some configurations are not statistically significant. These configurations correspond to the hypothesis of considering both fragment A1 and fragment A2 given to the lab in Arizona (Fig. 4.3).

According to this statistical analysis therefore it is unlikely that sample A2 has been dated. Well, the Tucson lab director Timothy Jull confirmed to Prof. Fanti that actually only sample A1 had been used for the radiocarbon dating, thus indirectly verifying also the reliability of the robust statistical analysis performed on the Shroud samples.

Concluding the presentation of the results regarding this controversial radiocarbon test performed in 1988, it can be stated that:

(1) Results obtained are not statistically reliable because of incorrect data evaluation.
(2) Environmental bias that could have altered the results even of thousands of years has not been taken into consideration.
(3) Historical events, especially in the first millennium, that could have influenced the Shroud, as far as environmental conditions are concerned, are not known. For example, the thymol exposition has perhaps influenced the chemical conditions of the fabric cellulose; who can exclude that during the earliest centuries the relic had been conserved in an aggressive environment like that with thymol of 1988, with the purpose of better conserving the Holy Linen?

(4) The little piece of fabric taken from a corner of the Shroud is not representative of the whole sheet in terms of dating.

(5) Not all the fundamental hypotheses of the radiocarbon dating method are verified; condition number 3 stating that there must be no environmental contamination is not verifiable.

(6) Condition number 4 regarding the sample representativeness is not verified.

(7) The deviation of the obtained result can also be caused by an environmental effect linked to the body image formation. The body image formation process is not yet clear, but it would seem related to an intense burst of energy.

(8) Since the process described in point 7 is not yet known to science, one needs to wait for clarifying the phenomenon itself before repeating a further radiocarbon dating test on the Shroud.

(9) The radiocarbon average date obtained by the 1988 test, of 1325 A.D., can be temporarily acceptable, provided that an uncertainty of some millenniums will be assigned instead of the ± 65 years declared in the report published in *Nature*.

(10) It is necessary to proceed with alternative dating methods, and this is will be the subject of Chapters 6 and 7.

Chapter 5

Journey of a Flax Thread

Knowing the identikit of the Shroud developed in Chapter 1 is not sufficient. The peculiar object under investigation has still many aspects to reveal, if we proceed with the eyes of a forensic detective to a more detailed analysis on the sheet (image, blood sampling) and also "into" the Shroud at the microscopic level (threads and linen fibers, their structure and origin).

While in the previous chapters we provided clues and traces of the Shroud in history as they emerged from historical documents (Chapter 1), from iconography (Chapter 2), and from a recent numismatic study (Chapter 3), in this chapter the material of the relic, namely flax, is under observation. We will carry out three exploratory journeys regarding flax.

The first is a journey that starting from the flax plant, will lead the reader within its stem in order to show the origin of the raw materials, which are textile fibers. Then, if the in-depth analyses in Sections A.4 and A.5 of the appendix are followed by the reader, in which the structures of a single fiber are described, the size scale of a millionth of a millimeter (40 billionth of an inch) can even be reached.[1] The materials constituting the fibers, especially their

---

[1]It is not easy for everyone to imagine how small a millionth of a millimeter (nanometer) is. Considering that a millimeter (0.40 in.) is equal to the distance

The Shroud of Turin: First Century after Christ!
Giulio Fanti and Pierandrea Malfi
Copyright © 2015 Pan Stanford Publishing Pte. Ltd.
ISBN 978-981-4669-12-2 (Hardcover), 978-981-4669-13-9 (eBook)
www.panstanford.com

particular microscopical organization, play a very important role in the alternative dating methods described in Chapters 6 and 7, since they use flax fibers coming from various textile samples of different ages.

In the second journey we briefly illustrate the process that, starting from the stem of the flax plant, leads to extraction of the textile fibers to obtain a thread and to realize a fabric. The processing steps may differ and be carried out by different means, depending on the historical period, so what we will propose here is a summary description to give the reader an idea of the complexity that characterizes the flax processing method.

The third and last journey describes and explains an interesting feature of the fibers taken from the relic: they are curiously distinguishable from the flax fibers coming from other earlier textiles. In fact many samples have been collected from the Shroud at different times and then subjected to various analyses in many scientific fields. The argument is developed in Chapter 9, while here, after a brief summary about the methods of collecting the samples, we will only speak about the flax fiber belonging to the Shroud.

5.1 Journey into the Flax Plant

The scientific name (by Linnaeus) of the common flax plant is *Linum usitatissimum*. As pointed out by the Latin adjective *usitatissimum* (*most useful*) this plant has been very much used since ancient times. In Turkey we can find traces of the use of flax for fabrics dating back 9000 years ago [94] and in Egypt at least 7000 years ago, and thanks to some rare archaeological findings, we have evidence that the production of ropes and other simple objects in linen goes back even 30,000 years ago [42].

The flax plant has several species (as many as 180) and it is now cultivated [7], as in the past, not only for the production of fabrics, but also to exploit the seeds from which flax flour and linseed oil are

---

between two consecutive notches of a common ruler, to arrive at one millionth of a millimeter we have to divide this distance into one million equal parts and then take only one of them.

derived; the linen is an annual plant that grows with greater ease in the temperate zones of the earth.

At maturity the greater varieties with textile interest reach an average height of about 1 m (3.28 ft.), presenting ramifications less marked in the upper part of the stem, at the end of which the flowers (blue) are placed and then the seeds grow. In general, referring to the cultivation of linen in Northern Europe, the sowing is done at the beginning of March and the maturation of the plant is completed in July [99].

Things are actually more complex because the so-called life cycle of the flax plant, which has 12 growth stages, is strongly linked to the climate of the growing region and also to the weather conditions of each year, which implies that the transition from one stage to another is not the same, neither everywhere, nor year after year in the same place.

However, without going into details, we can summarize the life cycle of the flax plant as characterized by [134]:

- a vegetative period (growth of the stem), which is performed in 45–60 days;
- a flowering period, which takes 15–25 days; and
- a maturation period of 30–40 days.

So the life cycle lasts at least 90 days.

Contrary to what we might imagine, to obtain flax yarn the whole plant is not used, but only the particular structures located in the outer part of its stem, which has a diameter of a few millimeters (around 0.1 in.).

If we make a cross section of a flax plant stem, among the various structures present in it (cuticle, epidermis, phloem or liber, xylem, etc.) and shown in Fig. 5.1, those with textile interest are located in the liber. Here there are about 30 bundles [35], with 10–40 fibers each, which extend parallel to the axis roughly for the entire stem of the plant. The single fibers forming the bundles are in fact the raw material to be extracted from the plant, because only these structures can be spun, namely transformed into yarn, after an appropriate technique of extraction and processing, which is explained in Section 5.2.

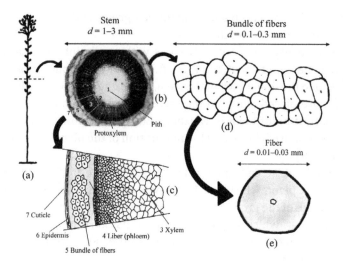

Figure 5.1 (a) Schematic figure of a flax plant. (b) A photograph of a section of the stem. (c) An enlargement of a portion of (b), and it shows some structures that are part of the stem. The structures of interest in textiles, namely the fibers' bundles, are situated at the perimeter of the stem and precisely in the phloem, also called liber. Each of the approximately 30 bundles (d) consists of 10 to 40 fibers with a polygonal section (e), with dimensions of the order of a hundredth of a millimeter, so 100 times smaller than the stem.

A single flax fiber has a variable length from 3 mm to 80 mm (0.118 in to 3.15 in.), while a bundle has a length between 150 mm and 900 mm (5.90 in and 35.4 in.).[2] Compared to fibers of other plants, those of flax are among the longest. Inside a bundle each single fiber overlaps with the adjacent ones over long stretches, so all the fibers are bound together (Fig. 5.2).

From a more technical point of view, the flax fibers are classified as *natural vegetable textile fibers of liberian type*. In fact [94]:

- they are *natural fibers* because they are already present in nature in the form of filaments, unlike the so-called technofibers, which are produced by humans from natural (lanital, rayon) or synthetic (nylon, kevlar) substances;

---

[2]The ranges given above are the result of the union of the values reported in the following papers: [18, 38, 39, 80, 139, 142].

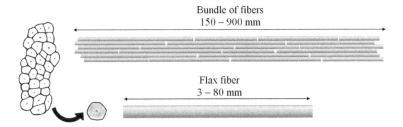

Figure 5.2 Schematic representation of the longitudinal arrangement of flax fibers within a bundle: the fibers overlap over long stretches with the neighboring ones and are bound together. Note the average size of a bundle with respect to the one of a single flax fiber.

- they are *vegetable fibers* because they are extracted from plants, but there are also natural textile fibers of animal[3] and mineral[4] origin; and
- finally they are fibers of the *liberian type* because they come from the phloem, also called liber (from which comes the term "liberian"), present inside the plant stem. Other bast fibers are, for example, those deriving from hemp, jute, kenaf, and ramie. The vegetable textile fibers, however, may also come from different structures of plants [131]: the leaves (banana, sisal, agave, abaca), the fruit (coconut), and the seed (cotton). Another type in this classification is the grass and reed fibers (rice, corn, bamboo).

Considering a single flax fiber, of which material is it made? In Table 5.1 we report the average chemical composition[5] of flax fibers

---

[3] For example, the wool (made from the fleece of animals such as sheep, goats, camels), the silk (produced by the serigene glands of the silkworm), or the "byssus" (a filament produced by the *Pinna nobilis*, a mollusk, nowadays very rare).

[4] We are speaking about asbestos, a set of minerals belonging to the silicate group, whose processing and production is nowadays widely banned in the world, since its high toxicity had been proved as a result of the diseases arising from inhalation of its fibers. For completeness, here we also mention another fiber category, the metallic ones, obtained with skill processing from precious metals like gold and silver by reducing them into long, thin strips.

[5] The reported values are percentages by weight; thus, for example, a value of 70% for a certain constituent indicates that in a mass fibers equal to 100 weight units, 70 are due to that constituent.

Table 5.1 Average composition of flax fibers and of other vegetable fibers by making a summary of the data available in the literature

	Cellulose [%]	Hemicellulose [%]	Lignin [%]	Pectin [%]	Wax [%]
Flax	64–75	11–20	2–5	1.8–3	1.5
Cotton	83–93	5–6	–	1–5	0.6
Hemp	68–78	15–19	3–11	0–3	0–0.8
Jute	60–72	12–22	5–16	0.2–1	0.5

References of the data: [18, 19, 25, 65, 127, 131, 136, 139, 142].

compared with those of same other vegetable fibers as it turns out from a synthesis of the available data in the literature.

We should note that the variations shown by the data in Table 5.1 are due not so much to the "inaccuracy" of the technical analyses, but rather and above all to the natural variations in composition from plant to plant and between the fibers arising from different areas of the same stem. In fact vegetable fibers are generated by a living organism whose growth, as we said, is influenced by various environmental factors, and it is performed with different growth rates into the structures, depending on the cultivation place.

From Table 5.1 we can deduce that the principal constituent of all natural fibers coming from plants is cellulose, a fundamental constituent composed only of the elements carbon, hydrogen, and oxygen. Cellulose is structurally formed by long chains without ramifications for unifying many basic building blocks (more information about the cellulose structure and characteristics are summarized in Section A.4 of the appendix). In chemical language cellulose is a polymer,[6] more specifically a polysaccharide.[7]

---

[6] This word with a Greek origin literally means "that has many parts," In fact a polymer is the result of the union of many basic building blocks (i.e., the parts) forming chains. In Section A.4 of the appendix the nature of cellulose's building blocks is explained also by means of a schematic drawing (Fig. A.1).

[7] The meaning of this Greek word is "that has a lot of sugars," and it is used to describe molecules formed by the union of many sugars. In the cellulose the "fundamental brick" is a molecule composed of two identical sugars bound together in a particular manner. More technical details are in Section A.4 of the appendix and in the schematic drawing in Fig. A.1.

The cellulose chains are also able to arrange themselves parallel to each other for long sections in a very ordered and compact way (see Fig. A.2 in the appendix), giving rise to a three-dimensional configuration that can be very tidy and imagined similar to that of noodles inside their packaging. This type of cellulose with very ordered chains is known as crystalline cellulose compared to amorphous cellulose, which presents chains arranged in a more disordered way. Crystalline cellulose plays an important role in the flax mechanical alternative dating method (Chapter 7), because over time cellulose tends to change from crystalline to amorphous, thus modifying the mechanical properties of the linen fibers.

Hemicellulose, lignin, pectin, and other minor constituents such as waxes (Table 5.1) follow cellulose as the fundamental constituent of all vegetable fibers. The reader may consult Section A.5 of the appendix for a synthetic description regarding the structure and salient features of the other substances mentioned in Table 5.1.

Let us now focus on a single flax fiber. We must warn that what follows, although interesting in relation to the internal constitution of a single flax fiber, may not be easy to understand for everyone because of its more technical nature. So readers who encounter difficulties can go directly to the next Section 5.2 "From the flax plant to tissue."[8]

Regarding a single flax fiber, a first important aspect to highlight is that it does not have a circular cross section but an irregular polygonal section with a number of sides [18] ranging from five to seven. Therefore to use the word "diameter" to describe a flax fiber would not be correct, even if this term is still widely used in the literature; it describes the cross section of the hypothetical cylinder that best approximates the vegetable fibers. So when the term "diameter" is used, a fiber's model with a constant circular cross section is assumed. In addition we observe that the diameter of

---

[8] For those who wish to leave the section at this point it is sufficient to know these two things: The first is that the flax fiber's surface is surrounded by a thin layer called P1, where there are irregularities (defects) called "kink bands," which occur as bulges like those of the bamboo cane; the second is that the underlying layer indicated with the abbreviation P2 is the thickest, and it is the one that most influences the mechanical properties of the flax fiber, in particular its stiffness and tensile strength.

a single flax fiber is not constant: it is variable along its length [35], and it grows [39] from the top to the bottom of the stem.

Flax fibers have average diameters [18] ranging between 0.005 mm and 0.076 mm[9] (0.0002 in and 0.003 in.), but with more frequent values between 0.007 mm and 0.025 mm (0.0027 in and 0.0010 in.), as we derived from the measurements of the fibers used in the mechanical alternative dating method.[10] Regarding the Shroud, we remember that the flax fibers analyzed of the relic have an average diameter of about 0.013 mm (0.005 in.), but we can find fibers with a diameter larger than 0.020 mm (0.0008 in.) or less than 0.007 mm (0.0003 in.).

The internal structure of a flax fiber is complex and fascinating at the same time because it is organized in concentric layers, like those inside an onion, which are described in Section A.6 of the appendix.

Among the various concentric layers the thickest (which can rise to about 80% [139] of the fiber's cross section) is the one named in the literature with the acronym "S2" (which stands for "secondary cell wall 2"). This layer is rich in microfibrils , spirally wrapped [18] with an angle of $10°$ with respect to the axis of the fiber, in which the crystallinity level can even reach 90% [131]. It is this underlayer that most influences the mechanical properties of the flax fiber, in particular its stiffness and tensile strength (namely when the fiber is stretched parallel to its axis).

Imagining a fiber, it is logical to represent it as perfect, but a real one is not at all. Fibers have often cell wall defects that reduce the tensile strength. Depending on their orientation with respect to the axis, the fibers' defects can be [10] longitudinal or transverse (Fig. 5.3); the transverse defects are known with the names of "nodes," "dislocations," or "kink bands". In some areas of the flax fiber's surface of the P1 layer during the growth phase the ordered arrangement of the microfibrils is somehow impeded, which spawns the defects, which occur as bulges like those of the bamboo cane.

---

[9] On the systematic use in the text of the millimeter unit instead of other subunit, see footnote 7 on p. 9.

[10] The details of the method are described in Chapter 7. The presence of dispersion in values is still linked to the fact that the fibers are generated by a living being, the growth of which involves the construction of various layered structures similar in architecture from plant to plant and within the same plant but never exactly alike.

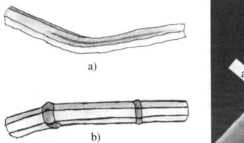

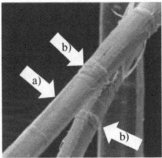

Figure 5.3 On the left, schematic representations of the defects in the flax fibers: (a) example of a longitudinal defect, which occurs as a groove on the surface of the fiber; (b) example of transverse defects (known as "nodes" or "kink band"), which appear as bumps surrounding the fiber. On the right, photograph of flax fibers by electron microscopy: the arrows indicate the two types of defects as they occur in reality.

Figure 5.4 Drawing showing how the transverse defects appear by observing a flax fiber under an optical microscope.

The larger transverse defects are visible under an optical microscope both under natural light and under particular lighting conditions (polarized light[11]). They appear as superficial lines crossing the fiber and are oriented roughly perpendicular to the longitudinal axis. Figure 5.4 shows a schematic representation of the transverse defects as they appear when observing a flax fiber with an optical microscope. In Fig. 5.10 are transverse defects observed under the microscope in cross-polarized light in the case of recent fibers and in those coming from the Shroud.

The presence of defects is invoked to explain the following behavior of the fibers: pulling a fiber like a rope, the fracture zone does not always occur in correspondence with the minimum diameter (namely the narrowest cross section of the fiber) as would be logical, but only with a probability [79] of roughly 40%–60%. The

---

[11]It is light with particular oscillatory characteristics explained in Section A.8 of the appendix.

defects are in fact potential weak points for the fibers when under stress.

It has been experimentally observed [10, 131] that the defects are incremented by the extraction phase from the stem and by the fibers' processing, which are discussed in the next section. The defects cannot only switch the failure zone to a different cross section with respect to the narrowest one, but in certain cases they can also heavily impair the mechanical strength of the fiber.

The journey into the flax plant ends here; the information provided, although sometimes simplified, is considered sufficient to make it possible to understand some aspects discussed in Chapters 6 and 7.

It is possible that most readers may be a bit surprised by such an unexpected complexity within a flax fiber from the microscopic point of view. A linen fiber is undoubtedly very complex, being a multilayer composite material with a fascinating architecture, just only thinking about the hierarchical organization of its internal structures (Section A.6). If what is inside a flax fiber appears complex, this is actually nothing compared to the complexity involved in the Shroud cloth and its body image!

5.2 From the Flax Plant to Tissue

The previous journey, into the flax plant, showed that the raw material (the fibers) is gathered in bundles within the phloem (liber), a peripheral part of the stem (Fig. 5.1). In this journey we illustrate, even if in a simplified manner, how the fibers are extracted from the stem and separated from bundles, how they are processed to get yarn, and how 2000 years ago linen fabric was produced.

Once the flax plants have reached maturity, the operations that can be performed in order to obtain linen yarn are several; in fact they do not need to be all done and least of all in the same sequence. However, over the centuries, some techniques have been abandoned in favor of others or improved in execution. So, since there are alterations, here we illustrate a possible sequence of operations to give a broad idea of the production process.

At maturity the flax plants are harvested. This operation, which is common in a huge variety of plants cultivated by humans, in the case of flax for textile use is realized more, than by cutting the stem, by pulling up the plant by the roots because [134] this contributes to obtaining fibers of better quality. The manual pulling, also mentioned in *Naturalis Historia*[12] (Natural History) by Pliny the Elder, in the past was clearly much more labor intensive, while it is now completely mechanized.

Then retting follows, an important process on which [108] the final mechanical quality of the flax fibers highly depends. The procedure[13] is performed by leaving [99, 108] the harvested plants on the field in swaths from three weeks to a month and a half. In these conditions the stems are naturally colonized by fungi that release enzymes. The enzymes attack and disrupt the pectin polymeric chains into the middle lamella layer (see Section A.6 of the appendix), with the result that the single fibers are more easily separable from the phloem and from each other within their bundle.

It is for this action that the retting phase has great importance in the production of flax fibers and the control of the degree of retting is a crucial operation. As a result an insufficient enzymatic attack would increase the difficulty of the bundles to detach from the structures of the stem within which they are immersed; hence the fibers would be characterized by a lower average quality because they were most damaged during the extraction process. On the other hand, as observed by N. Martin [99], an enzymatic attack that was too prolonged can also involve the fibers' layers, thereby reducing their mechanical strength.

The development of any natural retting, like dew retting, is not easily controllable, since both the environmental temperature and

---

[12]Book XIX, Chapter III, Paragraph 16: *Tum evolsum et in fasciculos manuales colligatum ... (Then pulled up by the roots, made up into small sheaves that will just fill the hand ...)*. The work, a naturalistic treatise in 37 books, was published in 77 A.D.

[13]Since there are several retting processes, the one described here, is the so-called dew retting. In fact it is possible to realize a retting process for the flax plants [134] before harvest (stand retting) or after harvest but following a different technique than the dew retting, for example, by dipping the stems in water into tanks (water retting) or by using chemical retting in which stems are introduced into special process tanks (reactors) with water and suitable substances (enzymes) in addition.

the atmospheric humidity greatly influence the process, in addition to the mass of straw per field surface unit (density of the harvested plants).

The color taken by the stems seems to be [108] a good way to evaluate the degree of retting: in fact, in dew retting within a week the color of the stems turns from yellow/green to gray, and then the gray shade becomes gradually darker. In any case, a lot of experience is needed to identify the optimal degree of retting without going over.

After the retting process the steps to obtaining fibers from the dried stems follow. Pliny the Elder writes: *The stems [...] left to dry in the sun, then, once thoroughly dried, they are beaten with a tow-mallet on a stone.*[14] In a nutshell, gathered in small bundles the stems were beaten to drop the seeds, then they were crushed and shattered with proper tools[15] on a hard surface (like the one of a stone).

Before the introduction of special motorized or manual machines capable of extracting fibers faster and with greater efficiency from stems, the operation described, traditionally hand-made by women in the autumn, was very tiring.

In ancient times it might take several strokes of the stems because the technique used to extract the fibers was overall more rudimentary and less efficient, so much more laborious than the manual ones of a few centuries ago or those in which special machines are used. This fact is probably one of the reasons that justify a particular feature of the Shroud's fibers: being distinguishable, under special lighting conditions, from flax fibers coming from more recent fabrics! How this is possible will be explained in the next section.

Performed by special combs with nails of different fineness, "hackling" allows the separation of long and straighter fibers from those short or ruffled.[16] With this operation any extraneous matter

---

[14] The Latin text is *virgae ipsae [...] sole siccantur, mox arefactae in saxo tunduntur stuppario malleo*, Liber XIX, Chapter III, Paragraph 17.

[15] A tool utilized in the past for this purpose was a wooden spatula formed by a flat board provided with a handle, used on its edge side and called "scutch"; the relative phase of typing bunches of flax stalks was known as "scutching".

[16] Long and straight fibers were utilized for the production of high-quality fabrics, while the shorter ones, if spinnable, were used to make fabrics less valuable, such as wipers, bags or, as Pliny states, in raw form as wicks for oil lamps.

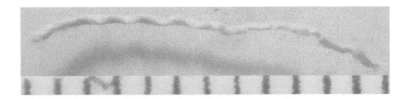

Figure 5.5 Linen yarn of the Shroud (the reference scale is in millimeters).

(shive) was also removed from the fibers. At the end of this process the so-called raw linen was obtained, in other words the flax fibers ready to be spun.

The yarn was finally obtained by spinning (Fig. 5.5). As G. M. Crowfoot states [42] in this phase the operations performed on the fibers are two: fibers are stretched in the sense of their length, thereby making them parallel as much as possible; in addition, fibers are pressed against each other by twisting and this makes them link together.[17]

Depending on the clockwise or counterclockwise twisting of the constituent fibers, yarns can have a "Z" or an "S" twist (Fig. 5.6). This particular notation is also used to characterize the direction of winding of wires in the strands and of the strands in a rope. If the twisting is clockwise, overlapping ideally the letter Z on the yarn, the underlying fibers perfectly follow the direction of the oblique segment of the consonant; otherwise, the winding is an "S" twist. As has been mentioned in Chapter 1 the yarn of the Shroud is a "Z" twist, as it is clearly evident on the right in Fig. 5.6.

The fibers in the relic's yarn can be seen very clearly with a simple optical microscope and we might count 80–200 fibers inside, as has already been noted in Chapter 1. If for all flax yarns, like those of the Shroud, the most evident substructure are the single fibers that form them, in a linen fabric or in any other realized with different material (silk, wool, hemp, cotton, etc.) we always pick out

---

[17]These two processes can be done simultaneously or in two successive stages, namely first stretching and then twisting. Rather well-known devices connected to spinning are the spindle and the spinning wheel, but it should be noted that we can spin without the use of tools, for example, by simply rolling the fibers between the palms of our hands.

Figure 5.6 On the left: Schematic drawing of the "Z" twist and the "S" twist yarn. On the right: Magnification of the Shroud cloth of Fig. 1.5 in which we clearly see the clockwise twisting ("Z" twist) of the flax fibers into yarn.

the yarn as the essential constituent substructure. But how was a linen fabric obtained 2000 years ago?

The answer is simple: A linen fabric, as well as the one of other tissues, is realized by weaving on a loom the weft thread through the warp yarns following appropriate rules (Fig. 1.5), depending on the strength, quality, refinement, and technical skills, clearly different in the various historical periods.

The warp is made up of many fixed and taut yarns, called passive. In the model the most primitive handloom used by the peoples of the Mediterranean during the Roman Empire and also after its fall, and probably used, with a bit more complex form, for weaving the Shroud (Fig. 5.7), the warp yarns were usually wrapped [110] on a log at the top and held taut with weights (pebbles or even realized in bronze).

In the manufacturing process of a linen fabric a spool with a continuous yarn, called "active," was moved between the warp threads, thus generating the weft.

To achieve this goal heddle frames were used (and are still used even in the so-called automatic looms), which are spacing rods of the warp yarns. In the reconstruction of an ancient Palestinian handloom shown in Fig. 5.7 there is only one heddle frame (the middle horizontal rod) that allows to appropriately split the weft yarns so as to obtain the simple one-over-one weaving type. In this case a square mesh fabric is produced in which one weft thread passes alternately over and under the various warp yarns.

In Chapter 1 we noted that the Shroud has a unique weave pattern with a herringbone band width about 11 mm (0.43 in.),

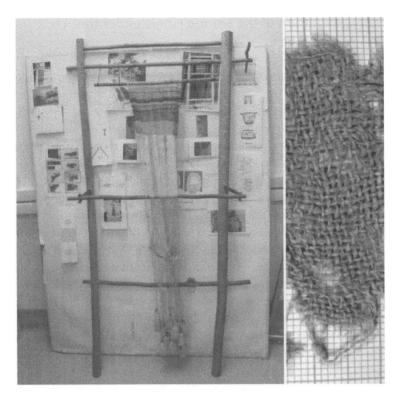

Figure 5.7 On the left, an example of reconstruction of a handloom used 2000 years ago in Palestine (Israel Antiquities Authorities, Jerusalem); on the right, an example of linen fabric with one-over-one texture from En Gedi (Israel), dated second century B.C.

which is the result of a diagonal weaving technique called three-over-one: each weft yarn is passed over three warp threads, under one, over three again, then under one (Fig. 1.5), and this sequence is repeatedly offset by a warp thread every next step.

The Shroud's more complex weaving technique, much more laborious compared to the simpler one-over-one, has required the use of a more complicated handloom with three heddle frames instead of just one (for further details see Section A.7 and Fig. A.5 in the appendix). The higher manufacturing complexity clearly increases the cost of the fabric: for this reason the Shroud cloth is considered highly prized for that time.

5.3 Recognition of the Shroud Fibers under the Microscope

For the media clamor which had the news, the best-known sample-taking operations from the Shroud's cloth were undoubtedly those performed on April 21, 1988, to allow the accomplishment of the radiocarbon dating of the relic with the ^{14}C method (Section 4.2); a small strip of the cloth's fabric corresponding to the corner of the frontal imprint, on the side of the left foot, was cut (Fig. 4.2).

Referring the reader to Chapter 4 for the details of this particular invasive sampling done on the Shroud, it is important to know that in reality other official far less invasive samplings from the relic have been made, not only in connection with the studies of 1988, but also going back to those performed by STURP[18] in 1978. Chapter 8 by M. Conca is an in-depth study about the various fragments coming out from the relic.

On the Shroud, among its flax yarns, among the fibers of its threads and in adherence to them, there are numerous hidden traces of its past, which can be studied through meticulous analyses. There were two techniques that have been used to collect these microtraces, most frequently named "dusts":

- The use of adhesive tapes, which were placed in contact with the surface of the Shroud to make fibers and powders adhere
- The use of a special vacuum cleaner equipped with appropriate filters, capable of retaining powders

Among the various more or less documented samples, already in 1534, during the restoration after the fire of Chambéry, Savoy gathered a sample of tissue from the Shroud. Moreover the Shroud's yarns taken in the past centuries are commercially available, although the trade in relics is considered sacrilegious for the Catholic Church.

---

[18] We recall here that it is the acronym of the Shroud of Turin Research Project. As we said at the end of Chapter 2 in 1978, 50 scientists carried reliefs and analyses on the Shroud's cloth for 120 consecutive hours in order to fulfill a multidisciplinary scientific survey on the relic.

Regarding the most recent samples, it must not be forgotten that in 1973 samples and crusts of blood were collected by the Pellegrino Technical Commission. Furthermore both in 1973 and 1978 dust samples were taken from the Shroud by the Swiss palynologist[19] M. Frei using adhesives tapes.

In 1978 a number of scholars, including the pathologist Pierluigi Baima Bollone and the STURP American chemist Raymond Rogers, took several samples, including image yarns and threads with crusts of blood (the first scholar), as well as a large amount of adhesive tapes placed in direct contact with the most important and representative zones of the Shroud cloth (the second scholar).

It was on this occasion that the technician Giovanni Riggi di Numana was authorized, among other things, to vacuum dust samples from the back of the Shroud, samples that have been widely used in the alternative dating methods described in Chapters 6 and 7. A particular head attached to a vacuum cleaner has been inserted, facing the Shroud's cloth, in the interspace between the sheet and the reinforcement Holland cloth, thus in correspondence of not visible areas. So the machine has vacuumed up the powder contained in this interspace (Fig. 5.8).

During the important analyses performed by STURP in 1978 and the cutting of the Shroud's fabric strip in 1988 further vacuum suctions were performed on the relic. In Chapter 9, we describe the type of material collected by the filters and some interesting findings resulting from the analyses. Here we develop a particularly curious characteristic concerning the flax fibers of the Shroud's yarns.

In the so-called dusts are present flax fibers arising both from the Shroud and from the Holland cloth (Section 2.3), which was acting as reinforcement and support at the time of these sample-taking operations. Since in the alternative dating methods discussed in Chapters 6 and 7 the Shroud's fibers taken from one of these filters (the "H" one) are used, the following question is more than logical: how have the relic fibers been recognized from those belonging to the Holland cloth? Only a few rows and the innovative technique used is soon revealed …

---

[19]See footnote 1 on p. 143 for the meaning of this term.

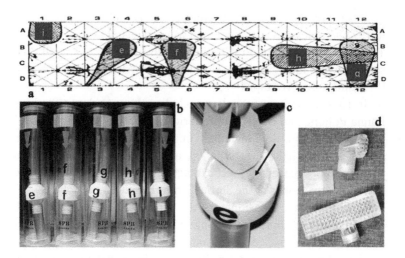

Figure 5.8 Dusts vacuumed from the Shroud. (a) Vacuumed areas named from "e" to "i"; (b) the corresponding filters containing the vacuumed dusts; (c) example of sampling of the filter "e" using a piece of adhesive tape; and (d) heads used with a vacuum cleaner to remove "dusts" from the back of the relic.

The first author had the possibility to directly analyze dust samples, fibers, and yarns taken from the Shroud (Fig. 5.9). Using these samples he developed a wide-ranging investigation involving different experts at the University of Padua and other Italian and foreign institutions. A grant of 54,000 euros ($68,000) to the first author from the University of Padua[20] and other minor ones (always provided by the university) have allowed him engage in research on the Shroud under several points of view, including one concerning the recognition of the Shroud's fibers from other textile fibers contained in the vacuumed dusts.

The research group at the University of Padua, led by the first author, with the collaboration of Roberto Basso, Irene Calliari, and Caterina Canovaro, carried out new investigations on the vacuumed dusts, which will be discussed in Chapter 9. Among the

---

[20] Research project of the University of Padua—Bids 2008, CPDA099244, Scientific Director Prof. Giulio Fanti, "Multidisciplinary analysis applied to the Shroud of Turin: a study of body image, of possible environmental pollution and microparticles characterizing the linen fabric."

Figure 5.9 Picture showing fragments of the Shroud received by the first author.

important discoveries in this field, there is the significant result of research carried out in collaboration with R. Rogers,[21] in which the following fascinating characteristic has been identified. The Shroud fibers are easily recognizable under an optical microscope if they are observed in cross-polarized light through an appropriate petrographic microscope.[22] In fact, observing the flax fibers by means of this instrument, they assume a special coloring. The technical reason why this happens is explained in Section A.8 of the appendix.

As can be seen in Fig. 5.10, while recent flax fibers show a multicolored coloring, but with a prevailing direction that is parallel to the fiber axis, those coming from the Shroud have a coloring quite

---

[21]This is the result of several email discussions with this scholar. In addition to a degree in chemistry, Rogers also obtained a diploma in microscopy from McCrone Institute.

[22]A petrographic microscope is similar to an optical microscope, but it has two polarizing filters, in addition, placed, respectively, before and after the sample and that can be rotated at will.

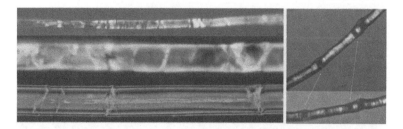

Figure 5.10 On the right, the particular coloring, like a coral snake's, of the Shroud's flax fiber observed under a petrographic microscope in doubly polarized light with two different rotations of the analyzer (the arrows indicate areas that correspond to each other). Let us note the presence of defects (knots) on the fiber. On the left, above and at the center, details of fibers, coming from the Shroud, seen in cross-polarized light; for comparison below there is a modern flax fiber. The average diameter of the fibers turned out to be 0.012 mm (0.0005 in.).

orthogonal to the fiber axis instead, comparable to the staining of a coral snake.

This is justified as follows. In agreement with J. Botella,[23] the particular coloring, similar to that shown by a coral snake, may be due to mechanical stress induced in the outer layer of the fiber during the process of extraction from the flax plant. While in the processing of recent flax fibers (after the first millennium) they are subjected to a relatively low stress level, in antiquity the processing by means of techniques less efficient and more invasive involves greater mechanical damages to the fibers. In the previous section we have already seen how in ancient times in order to extract the fibers the flax stems were more repeatedly beaten with sticks than in the modern techniques. Moreover Pliny writes:[24] [...] *after [the thread] is woven into a tissue, it is again beaten with heavy maces: indeed, the more roughly it is treated the better it is*. This means that mechanical stresses occur on the fibers in addition and so further damages are added when they are already part of the fabric. Besides it has been experimentally proved [10, 131] that the more heavy and numerous

---

[23] Shroud Science Communication of November 7, 2007. Botella is professor of biology at the University of Regensburg, Germany.
[24] The original Latin text is: [...] *textumque rursus tunditur clavis, semper iniuria melius*, Book XIX, Chapter III, Paragraph 17.

are the machining processes undergone by the fibers, the more the defects in the fibers increase.

Who might have ever thought that the damages to the Shroud's flax fibers would be so useful to recognize the fibers under a petrographic microscope? In this aspect the Shroud is amazing, but other characteristics even more surprising are yet to be revealed.

Chapter 6

Inquiries into Alternative Chemical Dating

In this chapter we will more intensely use scientific concepts, and it is not always possible to make them easily accessible to everyone. So the understanding of some terms might not be immediate to a reader unaccustomed to this kind of language. However, the dating results explained here are summarized with their main points in Chapter 9, to which the reader can go directly.

We have seen in Chapter 4 that the results of the 1988 radiocarbon dating carried out on the Shroud are not statistically and scientifically reliable, unless by assigning an uncertainty[1] of millennia to the obtained value but thus arriving to a result without historical significance. Consequently, we have highlighted the opportunity to develop alternative dating methods.

In this chapter we will discuss two chemical methods based on Fourier transform infrared (FTIR) and Raman vibrational[2]

---

[1] The concept of the uncertainty of a particular measurement has been described in Chapter 5, to which the reader can refer.

[2] From the point of view of chemical analysis, vibrational spectroscopy allows one, through appropriate diagrams, to identify molecules or characteristic groups in the molecules.

The Shroud of Turin: First Century after Christ!
Giulio Fanti and Pierandrea Malfi
Copyright © 2015 Pan Stanford Publishing Pte. Ltd.
ISBN 978-981-4669-12-2 (Hardcover), 978-981-4669-13-9 (eBook)
www.panstanford.com

spectroscopy.[3] But first of all, we think that it would be proper to start, even if in a simplified way, with some methodological considerations regarding what it means to develop alternative dating methods, since the matter is far from simple. In the following chapter, instead, we will deal with a mechanical multiparametric dating method designed by us with the acronym MMPDM.

6.1 Requirements of a New Dating Method

It is easy to say that as the radiocarbon dating method is not applicable to a certain historical sample, an alternative one must be used; what is less easy is using and, even more, devising an alternative dating method that could provide us with a scientifically reliable result. The question is indeed delicate and complex, and it involves several factors which are presented below.

First we must find a dating method that actually works and is sufficiently reliable, that is to say, which has an uncertainty level sufficiently reduced and it is as little as possible affected by the so-called bias[4] that can somehow corrupt the sample under testing and can consequently alter the result.

For example, a few years ago the Shroud of Turin Research Project's (STURP) chemist Raymond Rogers suggested [125] a chemical dating method of flax fibers on the basis of the determination of the vanillin content in the lignin[5] present in the Shroud fibers. Since a few samples were available to make comparisons and an in-depth study on the possible presence of environmental bias had not been carried out, Rogers was only able to make a "preliminary

---

[3]What is meant with spectroscopic methods is the study, and the consequent measurement, of a spectrum. Originally, a spectrum was the range of colors that could be observed when white light was dispersed into its components by means of a prism. With the discovery of the undulatory nature of light, the spectrum has come to be referred to as the intensity of light in relation to the wavelength or the frequency: this relationship is highlighted in appropriate Cartesian diagrams. The term "spectrum" has then been further generalized and was referred to as a flow or to an electromagnetic radiation intensity or to particles (atoms, molecules, and others) in relation to their energy, wavelength, frequency, or mass.

[4]The reader can go back to Chapter 5 for some references on the bias concept.

[5]The constituting materials of a flax fiber have been listed in Table 5.1. For some notes regarding lignin see Section A.5 of the appendix.

analysis" as is also reported in his published work. In this analysis he could only declare that the Shroud sample analyzed was "much older" than the date that the radiocarbon dating had established, but without being able to either provide a reference date or attribute an uncertainty level to the result.

The example reported above confirms what we have observed at the beginning: it is not easy to devise and to use alternative dating methods instead of relying on the more-than-tested ^{14}C (carbon 14) method.[6] In fact, the matter is very complex and delicate; according to the authors, it should begin well upstream of any good intuition that leads one to consider a particular physical or chemical characteristic as useful and, consequently, basing a new dating method on its measurements.

Therefore, in the light of the "Rogers case," the authors believe that it is necessary to start from the "general rules," namely to proceed in a more rigorous way from the metrological[7] point of view, by identifying and then by following few basic steps or essential "conditions" above, to obtain any dating alternative method to the radiocarbon dating one.

For the authors, there are 10 conditions that have to be satisfied, organized as follows from the procedural point of view:

(1) A property of the material under investigation that varies with time[8] must be found and defined.

(2) The relationship between the studied property and the time must be bijective, namely every value of the property must match one and only one value of the time elapsed from an initial reference instant.[9]

---

[6] For the chemical notation on isotopes see footnote 3 on p. 145.

[7] For the definition of metrology see footnote 2 on p. 144.

[8] Regarding the radiocarbon dating method the property is the measurement of the remaining quantity of ^{14}C.

[9] With bijective we mean a relationship between two elements belonging to two sets A and B: each element of A is linked to just one element of B, and each element of B is connected with just one element of A. For instance, the relation "capital city–European country" is a bijective relationship because each European state has only one capital city and after all each European capital city just belongs to one country. However, the relationship "biological mother–son" is not bijective: each individual child has a unique biological mother, but a mother can give birth to more than one son. Thus two different sons can have the same mother, because they are brothers.

(3) The relationship in question should not depend on other collateral factors. For example, the exposure of the sample to aggressive environments should not compromise the stability of the property that has been chosen, thus provoking undesired bias.

(4) A sufficiently large number of samples having known historical age must be at one's disposal in which it is possible to measure the property chosen to perform the dating method.

(5) By analyzing the results obtained from the above-mentioned series of samples, the method must be calibrated, defining the bijective sought relationship.

(6) To be scientifically acceptable, the procedure carried out until this point must find theoretical validations and confirmations.

(7) Once the bijective relationship has been defined (if it is possible in mathematical language and not only in diagrams), the uncertainty level must be associated to the relationship thorough opportune tests.

(8) Finally, the independence from other biases must be ascertained. If a bias influence turns out to be present, the property sensibility to it has to be measured.

(9) It is only at this point that it is possible to measure the chosen property with reference to the sample whose unknown date has to be brought to light. Then the corresponding date of the sample can be established through the bijective relationship found.

(10) The biases have to be considered in the light of the possible hostile environmental conditions, or, at least, of those conditions to which the sample has been exposed, that could modify the property under study. Therefore, the experimentally determined date must be corrected, and the corresponding uncertainty must then be assigned.

It is important to observe that once all these steps have been followed, it will probably not be easy to obtain the accuracy achieved by the radiocarbon dating method immediately, which has been tested and perfected through several decades. In other words, it is possible that at first the uncertainties concerning the dating carried out with new alternative methods are higher and could only be reduced in the incoming decades through an intense experimentation on a wide range of samples.

We point out that these 10 conditions are not easy to achieve. Even the well-tested radiocarbon dating method does not fully satisfy condition 2, because more than one value of the ^{14}C percentage is present in the defined calibration curve for certain periods of time. It follows that the resulting uncertainty turns out amplified if the measured ^{14}C percentage falls into those values in which the bijective relationship is not respected.

Moreover, even the radiocarbon dating method does not always respect condition 3, because the provision of recent carbon, like the one caused by bacteria interacting with the flax or that provoked by neutron radiations, alters the percentage of ^{14}C that is the aim of the measure, causing in some cases errors of even tens of millennia.

Despite these difficulties to cope with, the first author has sought ways to solve these problems that could be accepted for now and improved in the future through a more intense experimentation. First of all, a bibliographical analysis has been very useful to direct the course of the research toward certain properties instead of others. At the same time it has been necessary to try to better respect condition 4 in the research for samples of ancient flax fabric (Fig. 6.1 and Fig. 6.2).

After the preliminary selection performed on the samples to be subjected to the chemical vibrational analysis (FTIR/attenuated total reflection [ATR] and Raman described below), those effectively used for the determination of the calibration curves are reported from Table 6.1 to Table 6.3.

With reference to condition 5, the bijective relationship object of the research has to be determined according to an evident relationship between the values of the property under investigation and the corresponding historical dates. This precisely means "to calibrate the method." Yes, but how can it be achieved?

Let us suppose that we know the variation law of a certain property Y with time; for example, if Y decreases with the age of the sample following a linear trend, this means that in a diagram, having the years on the x axis and the values of the Y property on the y axis, what would result is a straight line. For instance, in the

Figure 6.1 Dr. Orit Shamir, director of the Israel Antiquities of Jerusalem, while she exposes some samples of antique fabrics to the first author for the choice of some fragments.

presence instead of a known exponential decay, like that of the ^{14}C method, we would obtain a negative exponential-type curve.

But when the variation law, namely the function $Y = f(x)$, is unknown, being still the object of the research, by measuring the Y property on a enough large number of samples of different ages, what we obtain is a set of points placed in a Cartesian diagram, having the years on the x axis and the values of the Y property on the y axis (Fig. 6.3). Due to various experimental reasons, also related to

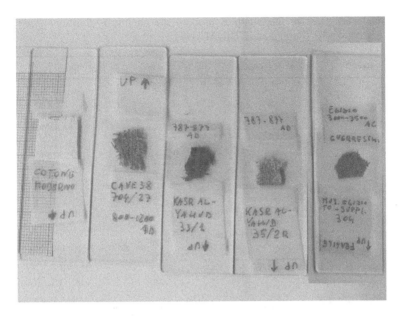

Figure 6.2 Some recent and ancient flax samples with a known dating method used for the calibration of alternative dating methods.

the repeatability[10] of the tests, the points reported in a Cartesian diagram are not "well ordered one by one."

What seems at first sight in a diagram to correspond to a linear or exponential trend must find experimental concreteness:

- by defining in mathematical language, according to a criterion, the actual variation law, that is, determine the function $Y = f(x)$; and
- by considering on the basis of objective—and not subjective—features the degree of its actual "description"

---

[10]Repeatability is the degree of concordance among the results of subsequent measurements of the same value carried out in such a way as to respect all the following conditions: same measurement method, same observer, same measurement instrument, same place, same conditions of use, and repetition within a short period of time. Such a characteristic is usually expressed in terms of the results' dispersion. If one of the above-mentioned conditions is not respected (e.g., the one concerning the same observer), we must talk of reproducibility and not of repeatability.

Table 6.1 List of flax samples used in the chemical vibrational spectroscopy tests (Raman and FTIR/ATR) (1)

Picture	Name	Description	Dating	Origin	Collection	Notes
	A	New herring-bone fabric 3–1	2000 A.D.	Liotex Italy	Private	Bleached with NaOH
	AR	New herring-bone fabric 3–1	2000 A.D.	Liotex Italy	Private	Similar to A for repeatability
	B	New fabric with rough texture	2000 A.D.	Graziano Italy	Private	Not blenched
	AII	Book paper	1800 A.D.	Italy	Private	Washed with water
	BII	Book paper	1746 A.D.	Italy	Private	Washed with water

with reference to the trend manifested by the points which are being measured.

As a matter of fact, we can draw infinite straight lines or exponential curves fitting the points reported in the diagram.[11] Thus we have to define the criterion with which we choose the "better" curve describing the trend of the Y property with time through a number of measured points. To characterize this aspect from a quantitative point of view, we have to refer to a suitable mathematical parameter capable of objectively evaluating the "goodness" of the choice that we have made. In other words, this mathematical parameter must permit us to state if a determined linear interpolating curve actually describes the data trend better

---

[11] In technical terms this operation is known as data interpolation.

Table 6.2 List of flax samples used in the chemical vibrational spectroscopy tests (Raman and FTIR/ATR) (2)

Picture	Name	Description	Dating	Origin	Collection	Notes
	DII	Medieval fabric 1–1	800–1200 A.D.	Pit 38 Qarantal Jericho Israel	Israel Antiquities Authority	Washed with water
	D	Medieval fabric 1–1	544-605 A.D.	Coptic Fayyum Egypt	M. Alonso	[14]C dated
	FII	Mummy fabric 1–1	55-74 A.D.	Masada	Israel Antiquities Authority	Washed with water
	E	Mummy fabric 1–1	300–500 B.C.	Egypt	Egyptian Museum Turin	Dated by historical information
	HII	Mummy fabric 1–1	1000–720 B.C.	Egypt	Egyptian Museum Turin	Dated by historical information

than others curves like, for instance, a quadratic, cubic, polynomial, or logarithmic one. So, to achieve this goal, we can use the mathematical method called least squares fitting based on the research of the so-called optimum curve able to minimize the deviations between the measured values and those that would be obtained if the mathematical model had perfectly described the relationship we were looking for. A valid method to judge the validity of the result obtained is based on the calculation of the so-called Pearson correlation coefficient.[12]

---

[12]The Pearson correlation coefficient is the measurement of the deviations between the mathematical model that has been hypothesized (straight line, exponential curve, etc.) and the pairs of values relating to a property under study and the corresponding historical date. Correlation coefficients inferior to 0.60 correspond to values that are too scattered in the interpolating curve determined with the least

Table 6.3 List of flax samples used in the chemical vibrational spectroscopy tests (Raman and FTIR/ATR) (3)

Picture	Name	Description	Dating	Origin	Collection	Notes
	K	Mummy fabric 1-1	2600–2200 B.C.	Egypt	Egyptian Museum Turin	Dated by historical information
	MII	Mummy fabric 1-1	2500–2200 B.C.	Thebes Egypt	Egyptian Museum Turin	Dated by historical information
	LII	Mummy fabric 1-1	3500–3000 B.C.	Egypt	Egyptian Museum Turin	Dated by historical information

With respect to condition 1, we studied different flax fabrics' properties, and in reference to condition 2 we defined various relationships between the properties chosen and the age of the fabrics. For example, it is known that flax turns yellow with time (see Section 1.2), and therefore its color has been considered as one of the possible properties to be correlated to the time in order to obtain the relationship needed to satisfy condition 2. However, it was found that condition 3 could not be verified because the coloring of the flax also depends on other factors, mostly environmental (temperature, humidity, and exposure to light), which can produce a bias that cannot be neglected and it is actually too influential to provide a dating method reliable enough.

In the case of the chemical composition of the cellulose, which can be measured through the FTIR/ATR and Raman spectroscopic methods, and in the case of the mechanical properties, it was

squares fitting method; values superior to 0.90 correspond instead to an excellent correlation, where 1.00 is the best value. For example, we will obtain the best value 1.00 by calculating the Pearson correlation coefficient of the points belonging to the same straight line. As a guarantee of their reliability, in these studies Pearson correlation coefficients superior to 0.90 have been found.

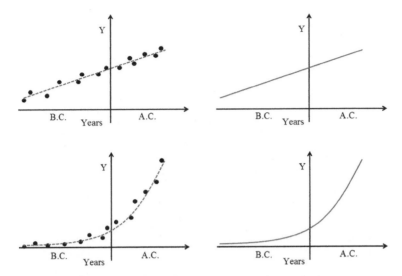

Figure 6.3 On the top, some measured values of a certain Y property changing with time reported in a Cartesian diagram (top left) and the corresponding diagram with a straight line interpolating the data (top right). On the bottom there is an example showing some points that follow an exponential decay law (bottom left) near (bottom right) the corresponding exponential interpolation curve.

possible to proceed to a dating of the fabric samples under study, even though condition 3 is not entirely verified.

Since the temperature is what mostly affects the sample, to check condition 3, we have carried out experiments on samples of recent Shroud-like linen, produced by the firm Liotex in the province of Turin, with the aim of assessing the effects of the more or less prolonged thermal exposure. Going into detail, to study the variations of the properties examined (Fig. 6.4), Dr. Stefano Dall'Acqua of the University of Padua has exposed Shroud-like linen samples in the oven for periods ranging from five minutes to five hours at temperatures from 150°C to 250°C.

In particular it has been noticed that the various environmental biases, like temperature and humidity, can alter, often heavily, the characteristics of the textile sample. However, these biases can be partially eliminated through an opportune preselection of the samples that have to be tested.

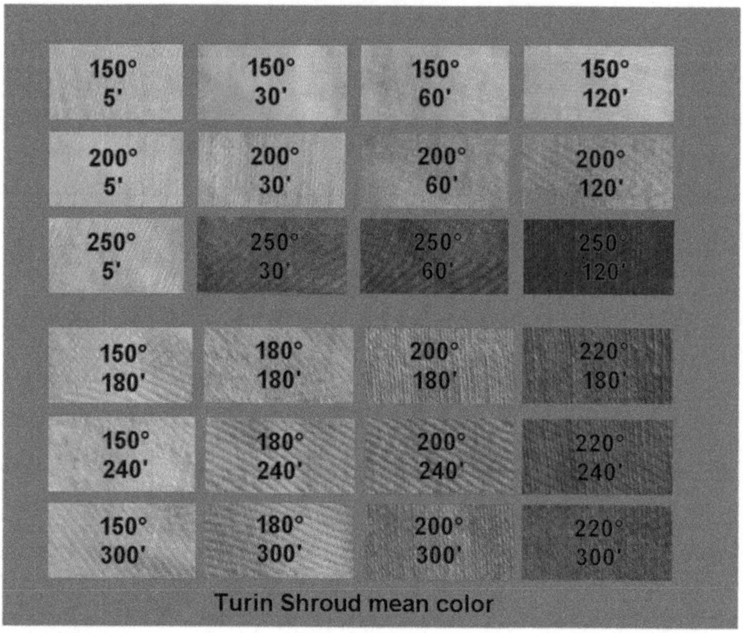

150° 5'	150° 30'	150° 60'	150° 120'
200° 5'	200° 30'	200° 60'	200° 120'
250° 5'	250° 30'	250° 60'	250° 120'
150° 180'	180° 180'	200° 180'	220° 180'
150° 240'	180° 240'	200° 240'	220° 240'
150° 300'	180° 300'	200° 300'	220° 300'

Turin Shroud mean color

Figure 6.4 Various coloring of Shroud-like linen fabrics exposed to heating for different times and temperatures.

With the purpose of reducing the influence of the possible environmental effects on the samples, before using them to define a possible bijective relationship between the properties and the date of the specimen, the textile samples conveniently underwent preventive optical (through a microscope) and mechanical analyses, which could eliminate the samples that were too contaminated. For instance, a textile sample coming from the Akeldama cave, in Jerusalem, was been immediately eliminated after preventive analyses, because it seemed to have been too altered by the humidity of the place where it had been exposed for centuries.

Even the soil of the place where the sample had been preserved can alter the result, most of all the spectrometric one. For example, in the case of the flax fabric (Fig. 6.5) coming from Masada (Israel, around A.D. 70) the initial analysis of the optical type at 100X showed the necessity to perform a preventive cleansing of the sample before subjecting it to the following analyses.

Figure 6.5   Sample of linen fabric from Masada (Israel), dating back to about A.D. 70. A preventive cleansing of the sample was determined by the initial optical analysis at 100X.

6.2 Dating Based on the FTIR/ATR Method

It is known from scientific studies carried out in recent decades that some photochemical properties of the cellulose contained in flax fibers can be highlighted through the so-called vibrational spectroscopy. This technique is based on FTIR and Raman analysis, which use a laser beam to excite the various energy levels of the molecules present in the sample under examination.

In the case of flax fibers, mainly made of polysaccharides, cellulose is the main component (Section 5.1). Such polysaccharides deteriorate over time, thus modifying their chemical structure. Among the various changes, here we report, as an example, the reduction over the centuries of both the cellulose degree of crystallinity[13] and degree of polymerization.[14]

---

[13] It is the ratio between the cellulose chains arranged in the crystalline state and those in the amorphous state.

[14] The degree of polymerization refers to the length of a chain of polysaccharide. In other words this parameter describes the average number of "fundamental bricks" (the cellobiose molecule in the case of cellulose, as can be seen in Section A.4 of the appendix) that are linked to each other in the polymeric chains. Time breaks the cellulose chains, making the fiber less resistant from the mechanical point of view.

Through the analysis of the chemical bonds, FTIR spectroscopy allows us to characterize the various materials and therefore also the polysaccharides, such as cellulose. But first of all, what is spectroscopy?

To understand this phenomenon, it is useful to provide a well-known example to everyone: the rainbow that is visible in the sky when the sun appears after the rain. The rainbow is characterized by seven colors, from red to violet, without any separation zone between each color. In physics, this characteristic is called the *visible spectrum*.

The rainbow is generated by sunlight beams passing through raindrops, which behave, in turn, as a set of prisms that separate the different colors of light, so forming the rainbow. A spectrometer such as FTIR works in the same way:

- A light source, in this case infrared, replaces the sun.
- The dispersive system, corresponding to the raindrops, is formed by the atoms of the material under analysis.
- The infrared sensor of the spectrometer has the same role as the eyes that see the rainbow.
- In the place of the brain, which perceives the rainbow's colors, there is a system that elaborates the information and provides a diagram of the spectrum corresponding to the substance under test.

Now, if we imagine that all the wavelengths of light have almost the same intensity, the corresponding spectrum of the rainbow given by the instrument would be formed by a continuous line practically horizontal; but if the sunlight were replaced by two monochromatic sources, red and yellow, for example, the spectrum would present only two peaks corresponding to the red and yellow wavelengths.

In the case of the FTIR spectrum, it is not continuous, but it shows characteristic peaks of the chemical substances present in the compound being examined. The analysis of the peaks of the spectrum allows us then to understand which chemical substances are present in the compound and the intensity of those peaks also allows us to know their amounts. Since the chemical compounds studied deteriorate over time, the analysis of the characteristics of

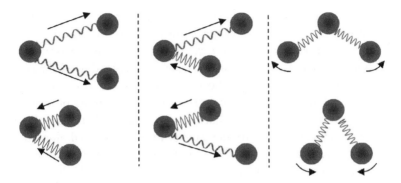

Figure 6.6 Examples of some characteristic vibrational motions of the atoms of molecules induced by the absorption of infrared radiation.

these peaks makes it possible to determine how much time has passed and thus to estimate the age of the fabric under examination.

In the machine an infrared beam hits the sample and its photons[15] are absorbed by the specimen's molecules, which reach in this manner an *excited vibrational state*.

To understand something more, we can conceive the atoms as many little balls linked together through springs in a molecule. If a photon strikes a ball, the atom starts to vibrate, transmitting the kinetic energy also to the other balls that are connected to it through springs. In this way a photon is able to generate a combination of relative vibrations among the atoms of a molecule (Fig. 6.6), each of which oscillates according to a particular frequency that can be observed through FTIR spectroscopy. It is by elaborating the particular oscillation frequencies shown by the atoms that the spectroscope is therefore able to identify certain atomic groups in the tested molecule and so determining its chemical composition.

When the molecules in an excited vibrational state return to their normal state, they emit a characteristic radiation that is acquired by the interferometer[16] of the instrument. Through a suitable

---

[15] A photon is the smallest quantity of electromagnetic energy.

[16] A typical interferometer consists of a moving mirror that, while turning, produces a difference in the optical path of the light beam: thus there is a constructive or destructive interference with the reflected reference ray.

mathematical operation called Fourier transform,[17] the acquired radiation is converted into an infrared spectrum (also called infrared diagram) in which the characteristic infrared vibrational frequencies of the substance analyzed are highlighted. This diagram is formed by two axes: the horizontal one is the abscissas, and the vertical one is the ordinates. The spectrum consists of a curve composed by several points. Each point corresponds the so-called wave number or spatial frequency[18] (which can be read along the abscissas' axis) and the radiation amount (which can be read on the ordinates' axis). The radiation amount is defined by the following two percentage parameters, absorbance and transmittance.[19]

We can try to better understand how Fourier transform works by thinking of spectators who attend a concert. They hear a sound produced by the different instruments of the orchestra, each of which emits a typical sound characterized by its own intensity. So the sound reaching the spectators' ears is composed of the audio frequencies emitted by a number of individual instruments. While the human ear has difficulty distinguishing the single sounds emitted by the orchestra, Fourier transform is instead able to highlight a spectrum in which all the audio frequencies emitted by the instruments during the concert are present, with their intensity in addition. Now, in the case of FTIR spectroscopy, the sound frequencies have to be substituted by the infrared frequencies emitted by the analyzed substances of which atoms or groups of atoms vibrate due to the excitation produced by the incident ray.

---

[17] Fourier transform is a mathematical operation based on the integral of the product of the studied function multiplied by an exponential function. It allows transforming of a signal from the time domain into the frequency domain. If, for example, we consider a sea wave, which can be approximated as a sine wave having a peak-to-peak length of 5 m (16.4 ft.), the corresponding Fourier transform presents a peak at a spatial frequency of 0.2 waves per meter ($1/5 = 0.2$).

[18] The spatial frequency corresponds to the number of cycles per centimeter completed by the vibration; its reciprocal corresponds to the wavelength of the ray considered.

[19] Absorbance (optical density in the past) is defined as the logarithm of the inverse of transmittance. Transmittance is the fraction of incident light at a given wavelength that passes through a sample. A transmittance value equal to 1 implies that in the passage through the sample, nothing of the incident radiation is lost (null absorbance), while a value of 0 implies that the radiation is unable to pass the sample because it has been completely absorbed.

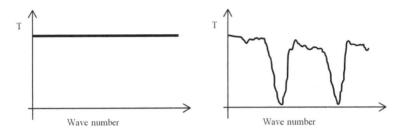

Figure 6.7 On the left: Example of an FTIR spectrum in transmittance of a material completely transparent to infrared radiation (i.e., whose molecules are not excited by the radiation itself). On the right: Example of a qualitative FTIR spectrum in transmittance of a substance not transparent to infrared radiation. The different absorptions of photons are shown in the diagram as a series of downward peaks of different intensity for each type of possible oscillation of the chemical bonds excitable by the incident infrared radiation.

Let us imagine, for example, that FTIR spectroscopy is performed on a material completely transparent to infrared radiation. Since its molecules would not be excited by the incident radiation, the resulting FTIR spectrum, with transmittance on the y axis, would turn out to be a horizontal line parallel to the abscissas' axis (Fig. 6.7).

In the most common case of material nontransparent to infrared, the various absorption of photons correspond instead to transitions of particular vibrational energy levels that are reproduced in the corresponding FTIR transmittance spectrum through a series of downward peaks of different intensity for each vibrational transition. Actually these are peaks that characterize the molecules contained in the material analyzed.

Professor Pietro Baraldi of the University of Modena, Professor Anna Tinti of the University of Bologna, and the first author carried out a study on the samples of the Shroud by using an experimental apparatus of the University of Bologna.

In particular, for this analysis the Nicolet 5700 instrument (Fig. 6.8) has been used. It is an instrument for FTIR/ATR spectroscopy, where the reflected beam by the sample under testing passes through an internal reflection optic element, which consists of a crystal having a high refractive index and placed in contact with the sample. If the incidence angle of the beam is greater

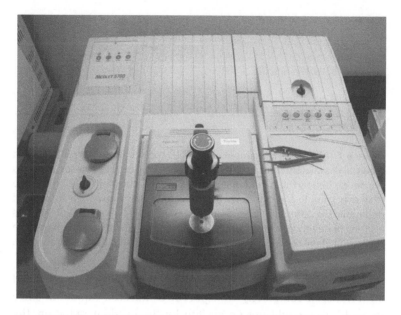

Figure 6.8 The Nicolet 5700 instrument used for FTIR/ATR dating of linen fabrics.

than a certain critical angle, there is total reflection. The ray then penetrates the surface of the sample around 0.5–2.0 thousandths of a millimeter (20–79 millionth of an inch) in thickness, thus producing a wave that is reflected several times on the crystal before reaching the sensor of the FTIR/ATR instrument, which will record the corresponding spectrum.[20] The FTIR/ATR spectrometer analyzes a sample's areas of 1 mm (0.039 in.), thus mediating every possible local discontinuity of the fabric (Fig. 6.9).

The new dating method by means of FTIR/ATR vibrational spectroscopy has been published [54] in the well-known journal *Vibrational Spectroscopy*.

Regarding condition 1, in this work three different ratios have been defined. The first one is that between the subtended area by the –OH group in the resulting spectrum and the subtended area by the C=O group. The second is the ratio between the subtended area

---

[20]Because of the absorption of the radiation by the sample, the reflected beam will be attenuated.

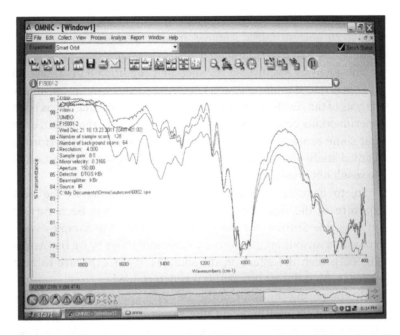

Figure 6.9 Example of an FTIR/ATR spectrum obtained from Shroud samples.

by the CH_2 compound and areas of the amorphous and crystalline cellulose regions. The third and last ratio is a combination of the compound C–O–C with the previous one just described.

The parameter that was better correlated to the historical date of the analyzed sample turned out to be the arithmetic mean of the three above-mentioned normalized ratios. The corresponding correlation coefficient[21] was very high, being equal to 0.956, thus indicating a close correlation between the chosen parameter and the historical date of the sample. This parameter is then suitable to represent the aging of the fabric on the basis of the concentration variations of the typical cellulose compounds and, more generally, of the polysaccharides.

With reference to condition 3, during the preselection it was observed that several fabric samples could not be analyzed following this method because they had been probably contaminated by dusts

---

[21]See footnote 12 on p. 193.

or colorants that altered the outcome spectrometrical result. Of the almost 20 ancient fabric samples initially prepared to be tested, around 30% had to be eliminated for this reason.

This analysis has instead revealed that condition 8 does have an influence, and therefore, it was necessary to carry out a sensibility analysis of the result in reference to the bias. In particular, it has been experimentally observed that temperature can alter the result, even after some centuries: for this reason it was necessary to perform a detailed study on the phenomenon in order to make a correction of the result obtained according to condition 10.

So, to investigate the sensibility of the method to the heat exposure of the flax fabric, FTIR/ATR analysis has been carried out on some Shroud-like linen fabric samples: the samples were exposed to various temperatures for different times by Dr. Dall'Acqua (Fig. 6.4); then the effect on the samples of the exposure to heat sources with different intensities was determined.

By comparing the color of the Shroud's linen with the samples exposed to heat and knowing that the relic was subjected to the Chambéry fire in 1532, it appeared that its fabric could not have been exposed to heat sources above 180°C for one hour or above 200°C for 30 minutes or even greater than 250°C for 5 minutes, because the part of the Shroud exposed to heat would be darkened and the effect would be clearly visible today.

In agreement with what had been reported in a recent publication, [58] on the basis of these experimental data we believed that it was necessary to correct the result obtained for the Shroud of +452 years.

According to conditions 9 and 10, the Shroud's samples were subjected to FTIR/ATR analysis. The combination of the three ratios considered for the dating gave a value of 0.387, which would correspond to the date of 752 B.C. ± 400 years at a confidence level of 95%, if the bias were negligible. However, since the results have to be corrected because of the effects of the exposure to the above-mentioned Chàmbery fire just noted above, we arrived at the Shroud dating from 300 B.C. ± 400 years with a 95% confidence level.

6.3 Dating Based on the Raman Method

The second vibrational spectroscopy dating method is Raman analysis, which allows studying of the molecular structure of the sample in relation to the behavior of the photons hitting the surface under test.

We owe to the Indian physicist Chandrasekhara Raman (1888–1970), who won the Nobel Prize in 1930, the discovery of the phenomenon. A laser beam can pass through a sample transparent to it or can be absorbed. In this second case, part of the ray is elastically scattered, while another minor part is scattered in an inelastic way.[22]

In a Raman spectrum the amplitudes of the different wavelengths of the beam scattered in an inelastic way are reported. These amplitudes can be related to the chemical composition of the sample. Since different molecules provide a particular inelastic reflection, the height of the spectrum's peaks is then proportional to the concentration of the substance generating each peak.

The ratios between the amplitudes of these peaks indicate the percentages of the substances in the sample under analysis; this characteristic has been used to determine the percentage decrement of particular molecular groups of the flax in relation to its degradation over time.

These tests have been carried out by A. Tinti and P. Baraldi in collaboration with the first author by utilizing the Brucker's MultiRAM instrument (Fig. 6.10), in which an infrared laser beam with a wavelength of 1064 nm[23] is used.

In reference to condition 3, for this kind of analysis a problem is caused by the fluorescence induced during the test, especially if flax or fabrics in general samples are considered.

Fluorescence is the property of some substances, such as flax, to emit electromagnetic radiation with different wavelengths after having been irradiated by a particular electromagnetic radiation. In

---

[22]The so-called elastic scattering of a laser beam consists of a reflection of the ray that preserves the same wavelength; the inelastic scattering instead produces a reflection having wavelengths different from that of the original beam.

[23]A nanometer is a millionth of a millimeter. A wavelength of 1064 nm corresponds to 4.19 millionths of an inch.

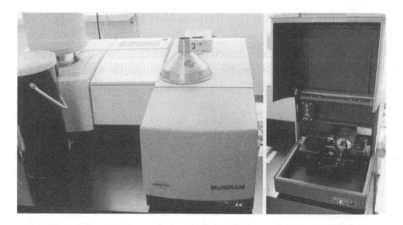

Figure 6.10 The MultiRAM instrument used for dating linen fabrics with the Raman method; on the right, particulars of the instrument's sensor where the sample is placed.

the case of Raman spectroscopy applied to flax fibers and threads, the fluorescence consists of producing an intense and wide-ranged set of electromagnetic radiation, which turns out to be added to the typical radiations characterizing the compound under investigation. As a consequence of this fact, the resulting spectrum shows the peaks of the analyzed substances mixed with those generated by the fluorescence, thus causing greater difficulty interpreting the obtained results.

It is to overcome this problem, instead of using a normal laser beam with a wavelength of 785 nm (3.1 millionths of an inch), which would amplify the fluorescence of the sample, that an infrared beam at 1064 nm has been used, because with this wavelength the fluorescence is considerably reduced.

Despite this, the obtained results were still affected by the fluorescence, which also varies from sample to sample. The calibration of the dating method through Raman analysis has therefore required the introduction of a third parameter, fluorescence, to be considered with the other two main parameters, the Raman amplitude and the historical date of the sample.

As for condition 1, the parameter chosen to be correlated with the age of the sample is the ratio between the intensities of the

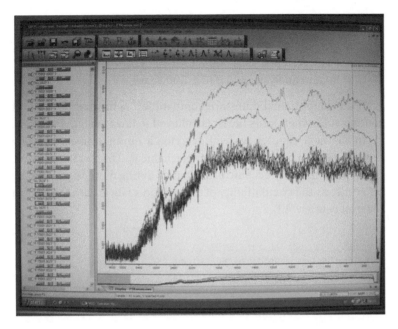

Figure 6.11 Example of Raman spectra obtained from the Shroud's samples.

following molecular groups' peaks: C–O–C and C–OH. As we said, to take into account the fluorescence of the flax samples, this ratio has been evaluated [54] in reference to two different values of fluorescence.

On the basis of conditions 9 and 10, the Shroud's samples were subjected to Raman analysis (Fig. 6.11). For example, in agreement with what has been reported in a recent publication [58], the ratio considered for dating has produced the value of 0.233.

We then obtained from this result that the calculated dating for the Shroud varies from 898 B.C. to A.D. 517. By approximating, and observing that is it highly improbable that the extreme values, related to the two calibration curves, occurred, we can affirm that the Raman dating of the Shroud's tested sample is 200 B.C. ± 500 years with a 95% confidence level.

Let us here summarize the two alternative dating methods used on the Shroud on the basis of vibrational spectroscopy analysis.

The two methods have provided the following dates at a 95% confidence level:

- FTIR/ATR analysis: 300 B.C. $\pm$ 400 years
- Raman analysis: 200 B.C. $\pm$ 500 years

Both the obtained dates are compatible[24] with that of the first century A.D., when Jesus Christ is believed to have lived in Palestine.

In the next chapter we will describe a multiparametric dating method that is not chemical but mechanical. The results that we obtained with this mechanical alternative dating method confirm by other means the compatibility of the Shroud's fabric with the date of the first century A.D.

---

[24] The term has been defined in footnote 10 on p. 138.

Chapter 7

The Mechanical Multiparametric Dating Method

The initial remark in Chapter 6 can be assumed as preliminary to this chapter.

While the two dating methods (alternative to the well-proven radiocarbon dating test) described in the former chapter are of chemical type, in what follows we depict a mechanical dating method for flax fabrics that we identified with the acronym MMPDM.[1]

In the following pages we synthesize the work [98] done from October 2011 to September 2012 by the second author for his master's degree in mechanical engineering. The degree thesis contains the design of a new machine for tensile tests on single textile fibers, the device calibration, tests on flax samples of different ages, data processing, and the way the MMPDM has been used for dating the Shroud.

As the reader can easily imagine, each step of this experimental work has various technical aspects, for example, critical questions regarding the machine's design and construction, metrological

---

[1] Mechanical multiparametric dating method.

The Shroud of Turin: First Century after Christ!
Giulio Fanti and Pierandrea Malfi
Copyright © 2015 Pan Stanford Publishing Pte. Ltd.
ISBN 978-981-4669-12-2 (Hardcover), 978-981-4669-13-9 (eBook)
www.panstanford.com

problems, estimations of the mechanical reference values, calculations justifying the abandonment of some constructive solutions, etc. Since our purpose is to write a nontechnical book, we decided to omit calculations and formulas in the text, preferring to describe what has been done and the main results that we have found in words and in a simplified manner.

At the end of the chapter, the reader will find that for the relic, the result given by the mechanical dating method turned out to be compatible with the first century A.D., the historical period in which Jesus lived. This confirms from the mechanical point of view what we have already obtained in a chemical way in the former chapter. The obtained result is summarized in Section 9.1 to which the reader less interested in the technical details can directly refer.

7.1 The Basic Idea

The passing of time virtually leaves its more or less tangible marks anywhere. The once bright colors lose their brightness over the years, the metal surfaces become dull, rust sooner or later attacks the iron manufactures, etc. If even stones are marked by time, we can assume that tissues are not exempt from this inevitable decay process. Among the damages caused by time in a fabric we decided to investigate the possible changes of its mechanical properties with aging.

In the case of flax fibers their polymeric substances can degrade over time. We have seen (Chapter 5) that polymeric substances[2] enter in different percentages in the cell walls and that cellulose is their main constituent (Table 5.1). The degree of polymerization[3] turned out to decline; thus the time tends to fragment the long chains of cellulose into smaller pieces. Consequently, the microfibrils[4] become more fragile. Another aspect, already mentioned in Section 6.2, is

---

[2] See footnote 6 on p. 168.
[3] See footnote 14 on p. 197.
[4] Microfibrils are thread-like agglomerations of cellulose molecules in crystalline form inside the flax fibers (Section A.4). They play the role of a reinforcing element.

the reduction of the fiber's degree of crystallinity[5] over time, which makes the polymeric chains in the microfibrils more mobile.

Some interesting experimental evidence has given strong support to the hypothesis that chemical degradation mechanisms regarding the fibers' structure due to aging are actually as much capable to modify the mechanical characteristics of a fiber as to let us, in agreement with condition 1 (Section 6.1), to use these properties for dating purposes.

At the Industrial Engineering Department of the University of Padua the beginning of the interest for the mechanical analysis of single fibers dates back to 2009. Knowing that at the Max Planck Institute (Germany) tensile tests[6] on such fibers were performed to determine the breaking load, namely the force value at which the rupture of the sample takes place.[7] The first author sent to this institute samples of linen fabrics belonging to different ages for evaluating such mechanical characteristics.[8]

The outcome was the following: In the specific case of the breaking load, a likely linkage, to be more deeply studied, emerged from the measured data between aging and mechanical characteristics' variation with the equipment available at the time. However, the institute was not able to provide measurable values of all the samples, especially for the older ones. In fact the instrument used for the force evaluation (the load cell) allowed it to appreciate only relatively high loads and the breaking load for some of the sent samples fell below the instrument resolution.[9] Therefore, even if

---

[5] See footnote 13 on p. 197.

[6] We will later describe (Section 7.4) the essential elements of a tensile machine. In principle, in such an experimental device a material specimen, grabbed in a suitable manner, is more and more pulled up until failure, recording certain parameters during the test.

[7] The tensile breaking load is a mechanical property of the materials. It quantifies the maximum force bearable by a specimen. Hence it is a parameter related to the materials' resistance; in fact below the breaking load the material has resistance capacity, because it does not break.

[8] Some samples are also included in Table 6.1 since they were afterward used in the chemical and mechanical dating methods.

[9] The resolution is a characteristic of a measurement instrument; it is the smallest change in a quantity being measured that causes a perceptible change in the corresponding indication (Section 4.14 of *International Vocabulary of Metrology, VIM*, III Edition, 2008). For example, with a common weigh scale we cannot measure

the data experimentally obtained were not sufficient to determine a relationship between the breaking load and the time, from those tests the evidence of the variation of that mechanical property with aging clearly appeared.

This outcome represented a very interesting starting point to try to develop a mechanical dating method for textile fibers alternative to the radiocarbon dating test. The new dating method is characterized by the following basic idea: if certain chemical structural modifications seem capable of changing the mechanical behavior of a flax fiber, then by measuring certain mechanical characteristics within a tensile test, some of them may be characterized by a bijective[10] relationship with time such as to be exploited for fabric dating.

Let us remember the basic conditions that should be met by any alternative radiocarbon dating method (Section 6.1).

As regard condition 1, referring to the experimental tests carried out by the Max Planck Institute in 2009, the breaking load appeared to be a promising mechanical property for the mechanical dating of a fabric. However, since different mechanical characteristics of a material can be in general determined with a tensile test, the possibility should also be considered that other properties may show variations with aging so as to be potentially exploitable for the construction of dating diagrams. We have used the expression "potentially exploited" because the identification of a distinct change with aging is not sufficient by itself; it is also necessary that this variation has a bijective behavior, as required by condition 2.

At the beginning of the work of the second author (October 2011), the challenge connected with the load cell resolution in measuring forces acting on ancient single flax fibers immediately raised the question whether purchasing a commercial tensile machine would be a feasible solution for conducting mechanical tests in the department laboratory.

---

the mass of a feather, because the mass variation produced by such an object on the instrument's measurement system is not detected, being too small. But if we instead set the same feather on the plate of an analytical balance, we can read its mass value: in fact the resolution of this kind of scale, even equal to 0.005 mg (0.00000018 oz.), is smaller than the mass of the feather and so its mass can be measured.

[10] See footnote 9 on p. 187.

7.2 Mechanical Aspects of the Fibers

Plant fibers are very much studied from the mechanical point of view. The cellulosic fibers, like the flax ones, are, in particular, giving promising results in automotive and aeronautical industries [137] as reinforcing elements [114] in substitution of the glass fibers.

There are several papers in the literature on both the microstructures inside the plant fibers[11] and the mechanical data deriving from tensile tests.[12] We focused our attention on the following three mechanical aspects:

7.2.1 *Type of Tested Specimens*

The type of specimens subjected to tensile tests are fibers directly extracted from a plant according to a process even faster than the one illustrated in Section 5.2.[13] The extraction of the fibers from the stems is manually performed with utmost care addressed to avoid damaging the portion of the fiber[14] to be tested by the tensile test.

Instead, we did not find papers in which the mechanical tests are conducted by extracting fibers from the threads of a fabric, because the use of these kind of fibers would increase the scattering of the results due to the presence in them of a larger number of defects (Fig. 5.3 and Fig. 5.4).[15] Nevertheless a dating method based on the mechanical behavior of flax textiles requires taking fibers from weaved threads.

7.2.2 *Mechanical Data of Textile Fibers*

The most frequently measured mechanical parameters, on which the technicians usually make comparisons between different textile fibers, are the following:

---

[11] For a technical deepening concerning flax see Section A.6 of the appendix.

[12] This particular mechanical testing method is known as the single-fiber tensile test (SFTT) [79].

[13] If the influence of the retting process is not itself the aim of the research, sometimes dew retting is replaced with other techniques (see footnote 13 on p. 173) in order to speed up the fibers' production.

[14] As we will illustrate (Section 7.4), to perform the tensile test the single fiber is usually glued onto a suitable support (tabbing technique). The gauge length of the fiber is the portion between the two drops of glue.

[15] We have seen in Chapter 5 that the fibers' processing in general increases the amount of damages in the fibers.

- *Percentage strain at break* is the ratio between the increment in length of the specimen at break and its original length multiplied by 100. It is a dimensionless quantity[16] here indicated by the symbol ϵ_R.

  If a material has $\epsilon_R = 2\%$, it means that just before the failure of the specimen its current length is 2% longer than the original one. According to the definition given above, this mechanical parameter quantifies the elastoplasticity[17] of the material.

- *Breaking strength*, indicated by the symbol σ_R, is the ratio of the maximum load recorded[18] to the initial cross section of the specimen. This mechanical characteristic is a measure of the maximum stress[19] that a material can withstand while being stretched before breaking.

- *Elastic modulus* (or Young's modulus) is a mechanical parameter that quantifies the elasticity of the materials, namely it measures the materials' resistance to being deformed elastically when a force is applied. We indicate the Young's modulus[20] by the symbol E.

  The following example explains the practical meaning of the elastic modulus. Let us imagine two perfectly identical-sized cantilever beams, the first made of steel and the second of aluminum. If we load these two beams with the same

---

[16] In fact, as the change in length is a difference between two lengths expressed in meters (or in inches), if we divide it by the initial length of the specimen (meters or in inches, too), the value that we obtain is actually dimensionless.

[17] In materials like steel, the percentage strain at break measures only the plastic deformation, which is evaluated by bringing together the two pieces of the specimen. The reader can refer to Section A.9 of the appendix for the concepts of elasticity and plasticity.

[18] Unlike steel, the maximum load coincides with the breaking force for the flax fibers.

[19] In the SI system the unit of stress is pascal (Pa). The correspondent US customary unit is pounds-force per square inch (psi). The breaking strength can also have the magnitude order of hundreds of millions of pascals. Consequently the megapascal unit (MPa) is normally used. A megapascal corresponds to one million pascals, for example, 400 MPa = 400,000,000 Pa.

[20] In the SI system the unit of the elastic modulus is Pa. The correspondent US customary unit is psi. The Young's modulus can also have the magnitude order of hundreds of billions of pascals. Consequently the gigapascal unit (GPa) is normally used. One GPa corresponds to one billion of pascals, for example, 205 GPa = 205,000,000,000 Pa.

Table 7.1 Values of the mechanical properties strain at break ϵ_R, breaking strength σ_R, and Young's modulus E for some vegetable textile fibers (flax's data are reported in bold in the first row). We also included for comparison the values of both E-glass and carbon nonnatural fibers

Fiber	ϵ_R [% *]	σ_R [MPa **]	E [GPa ***]
Flax	1.2–3.2	345–2000	12–85
Cotton	7.0–8.0	278–597	5.5–12.6
Jute	1.5–1.8	400–800	10–30
Hemp	1.6–3.5	310–900	30–90
Ramie	3.6–3.8	400–938	61.4–128
Sisal	2.0–7.0	511–700	9.0–38
Glass-E	*2.5*	*2000–3500*	*70*
Carbon	*1.4–1.8*	*4000*	*230–240*

References of the data: [18, 36, 37, 49, 90, 135].
* See footnote 16 on p. 214.
** See footnote 19 on p. 214.
*** See footnote 20 on p. 214.

downward force in correspondence of each of the free ends, we would observe that the aluminum cantilever beam turns out to be more flexed than the steel beam. In fact the Young's modulus of aluminum is smaller than that of steel; thus the higher is the elastic modulus of a material, while the smaller is the elastic deformation that occurs at the same loading, geometry, and constraint conditions.

Table 7.1 reports the values of the above-mentioned mechanical parameters for both some vegetable textile fibers and two nonnatural ones. The flax fibers have a Young's modulus comparable to that of the glass fibers and an average breaking strength greater than that of the other vegetable fibers, hence the interest in flax as a reinforcing ecological material replacing the glass fibers.

7.2.3 *Mechanical Behavior of Textile Fibers*

As regard the mechanical behavior of flax fibers, a clear scattering connotes the experimental values of the mechanical properties. As observed by V. Placet [114], this occurs because of the method used for both performing the tensile test and calculating the mechanical properties, but mainly because the fibers derive from a living being. There is also evidence that the fibers' processing increases the amount of defects, already naturally present (Section 5.1). The dispersion of the data also appears to be connected to environmental effects during the tensile test, and even to the "loading history," namely the fact that the fiber has already been previously pulled or not. In short, vegetable fibers have a rather complex mechanical behavior.

In developing an alternative dating method, the evaluation of all these aspects must be taken into account, as requested by condition 3 (Section 6.1).

Even if the strain at break ϵ_R, the breaking strength σ_R, and the Young's modulus E_f are the most measured mechanical properties of the natural textile fibers, there are other mechanical properties whose study could provide both a better understanding of the behavior of the single fibers and the development of most advanced models.

We point out that the analysis of the behavior of fibers subjected to cyclical loading–unloading conditions has been done [32] only with the aim of evaluating the variation of the Young's modulus cycle by cycle. But the internal energy dissipation of a fiber can also be studied through the cycles, for example, by calculating the loss factor,[21] as we have done (Section 7.7).

In the literature we did not find papers with this type of analyses on single flax fibers but this study can give birth to new promising interdisciplinary techniques of analysis.

---

[21] The loss factor is a mechanical parameter capable of quantifying macroscopically the many microscopic adjustments between the polymeric chains of a fiber (Section A.6) subjected to loading–unloading conditions. These internal adjustments dissipate energy. The loss factor is closely related to the damping, namely the property of a body to dissipate its internal energy under vibratory conditions. Let us imagine a guitar string. Once it is plucked, the string vibrates, but its vibratory motion spontaneously ends after a few moments. This occurs because the string internally dissipates a certain amount of energy, in addition to the energy dissipated by the friction with air. The greater is the damping, the more the energy is internally dissipated into the string and thus the faster the vibratory motion extinguishes.

7.3 Reduction of Other Environmental Effects

Over the centuries an ancient fabric may have suffered prolonged exposure to direct sunlight and/or to the oxygen in the air. Before being discovered, it may have been subjected to a cyclical heat condition related to the place (as in the case of a desert climate) or it may have come into contact with earth, sand, water, and/or various microorganisms. Furthermore, an old fabric can also have been stored incorrectly[22] for several decades after its discovery.

Our aim is to study the mechanical behavior of the single fibers coming from both modern and ancient flax fabrics in order to search for possible dependencies with time, to be used for the construction of dating diagrams. To achieve this goal, we need to assess whether the environmental conditions do have a bias[23] or not, namely whether they are able to affect the stability of the mechanical properties linked in a bijective relationship with time. This is in fact what condition 3 (Section 6.1) imposes among the 10 basic rules that should be met by any dating method, as well as by the radiocarbon one described in Chapter 4.

Hence the following question: Can the spontaneous and irreversible degradation of vegetable fabrics, which causes internal structural changes in the polymers forming the fibers, be additionally accelerated either by environmental factors [133], such as temperature, light, moisture, and soil acidity, or by attacks from lichens, molds, and mites, and rotting? In other words, do these environmental factors affect the mechanical results from an ancient fabric in such a way as to mask the possible relation between a particular mechanical property and the corresponding age?

It was found that systematic effects of an environmental type, first of all temperature and humidity, can also heavily alter the mechanical characteristics of a textile archaeological sample.

In particular, with regard to the temperature, we carried out tensile tests on fibers extracted from three series of the flax Shroud-

---

[22] For example, exposure to intense natural or artificial light sources, maintenance in a very humid environment, contamination with different kinds of gaseous, liquid, or solid substances, including those released by the container or produced by microorganisms.

[23] For the meaning of the term "bias" see Section 4.3.

like samples (Fig. 6.4) exposed in the oven[24] to different times and temperatures (Section 6.1). More precisely, we performed tensile tests on fibers always coming from the same fabric (the one identified with the name A in Table 6.1) but in these three different conditions: fibers not thermally treated (reference samples), fibers exposed in the oven for 120 minutes at 200°C, and fibers heated for 120 minutes at 250°C.

Even if the number of points experimentally obtained was not sufficient for evaluating interpolation curves with the least squares method (Section 6.1), the second author found that the thermal exposure has similar effects on the breaking strain, the Young's modulus, and the loss factor to those produced by time. So we can conclude that the exposure to a significant heat flow[25] is an event to be considered for a fabric. In fact heat degrades the mechanical response of the fabric's fibers, decreasing their stiffness, strength, and probably the degree of crystallinity.[26] Hence heat is able to produce a premature aging as a final overall result.

We obviously took this aspect into account when we used fibers from the Shroud for dating purpose by means of alternative methods to the radiocarbon dating one. The first author, who owns samples of the so-called dusts ("H" filter, Fig. 5.8), fragments (Fig. 5.9), and yarns (Fig. 5.5) coming from the relic,[27] also knows their origin location on the Shroud's sheet. The used samples were located far from the burned areas of the relic and they were most likely protected from heat fluxes due to fires, because the Shroud has been preserved for centuries folded into several parts and not outstretched as it is nowadays. Since flax is a poor heat conductor, it is easy to think that not all the layers have been subjected to extreme temperature variations.

The following aspect must also be considered. In Section 1.2 we described the yellowing of the fabric and in Section 6.2 we discussed

---

[24] This experiment was performed by Dr. S. Dall'Acqua of the University of Padua.

[25] Such as that coming from a fire, which is in effect an accidental exposure to an intense heat flow. We observed in Chapter 2 that the relic was in high danger of being destroyed by a fire in 1532.

[26] If the degree of crystallinity decreases, there is an increase of the amorphous zones and a corresponding decrease of the crystalline ones.

[27] For further information see Section 5.3, Chapter 8 by Marco Conca, and Chapter 9.

the effect of the heat of the Chambéry fire on the color of the Shroud's cloth. Let us recall here what we said there. By comparing the color of the Shroud's linen with the samples heated in oven we concluded that the analyzed samples from the Shroud could not have been directly exposed to a fire and even less to heat sources above 180°C for one hour, or above 200°C for 30 minutes, or even greater than 250°C for 5 minutes.

According to these considerations, outside the relic's burnt areas, the bias produced by the fires that have damaged the Shroud can be considered negligible, namely less than one century of uncertainty.

Moreover, the aleatory effects of the climatic temperature variations to which the relic's cloth has been subjected over the centuries can be considered not dissimilar to those undergone by the fabrics used for the calibration of the alternative dating method, and therefore they can be included within the overall uncertainty of the final result.

Humidity exposure is an environmental factor capable of influencing the mechanical results of an ancient fabric by embrittlement of its fibers in such a way as to alter its dating, namely making it apparently more ancient than it actually is.

This is the case of the already mentioned (Section 6.1) textile sample coming from the excavations of Akeldama (Jerusalem): During the tensile tests on single fibers extracted from its yarns, these fibers were very fragile, with a fragility comparable to, if not greater than, that of fibers of more than 4500 years ago. This was clearly incompatible with the dating, known, of the fabric. The abnormal mechanical behavior was immediately interpreted as due to an environmental alteration, particularly related to the humidity of the place to which it had been exposed for centuries. So the sample has been consequently discarded.[28]

There are two other aspects to be considered regarding the environmental bias effects in general. The first one concerns the effect of particular production techniques (specifically the bleaching with sodium hydroxide). The second aspect is related to the fact

---

[28]We add that at the first visual inspection the fabric in question was characterized by a blackish color. This fact, coupled to the abnormal fragility of its fibers, is further evidence of the heavy environmental alteration undergone by the fabric.

that the fibers to be tensile-tested come from a fabric's yarns and therefore they can be taken from different positions of the thread. Are these aspects capable of producing systematic effects to be appropriately considered in the development of a mechanical dating method? The answer is positive.

As for the bleaching treatment with sodium hydroxide (namely the exposure of the yarns' fibers to an aggressive environment), among the 11 series of selected fabrics (as described later) to calibrate the mechanical dating method, the two series A and B in Table 6.1 have this important difference. Sample A corresponds to a modern fabric bleached with sodium hydroxide, while sample B comes from a natural modern fabric. The two fabrics A and B differ for the manufacturing process and cannot both be used to represent the present in the dating method. In fact, for the samples A and B the comparison [61] of the values of the mechanical properties (breaking strength, elastic modulus, and loss factor) obtained from the tests carried out with our tensile machine (Section 7.6) revealed that the process undergone by sample A has changed its mechanical properties, making them worse. Since in antiquity linen fabrics were not bleached with aggressive compounds as it is done today, we excluded series A in the construction of the dating diagrams in order not to introduce systematic effects linked to a different manufacturing method. This is the reason why we considered only sample B, produced with natural techniques, in order to have more homogeneous samples of flax fabrics.

Let us now consider the extraction of the single fibers from the fabrics' threads. We can easily guess that the fibers closer to the center of a yarn are, on average, more protected from the environmental effects than those more superficial. On the basis of this consideration, we decided to pick up the single fibers always using the same procedure, taking them from the outside of the thread. We followed this procedure also because the Shroud's fibers of the so-called vacuumed dusts ("H" filter Fig. 5.8) derive from the outside of the threads.

According to the obtained experimental results, this procedure turned out to be a good choice. In fact we compared the data of tested fibers, picked up as indicated, with those taken from the middle of the thread. These latter fibers (i.e., less exposed) resulted to be "younger" [98] due to environmental effects with a bias even of the order of centuries.

What we highlighted shows that accuracy in the choice and preselection of the samples play a key role in the environmental bias containment. As a possible future evolution of the mechanical dating method, the determination of different dating curves as a function of the procedure adopted in picking up the fibers from the threads could be considered.

Although condition 3 is not completely verified, we were nevertheless able to mechanically date the Shroud by means of tensile tests on its fibers. We achieved this goal because, in agreement with what was suggested by W. Hu [79] in the case of modern fibers, we performed a proper preliminary selection of the sample to be used before conducting the tensile tests. In fact this operation limited the bias, thus reducing the uncertainty that accompanies the final result.

We now describe the adopted preliminary selection procedure. We used specimens coming from various fabrics of different ages. Each fiber (namely the sample), deriving from the external part of the threads, before being subjected to the tensile test, must pass:

(a) a visual test consisting of a microscopic analysis in the magnification range 10X–100X in which the fabric color is also considered;
(b) a visual test consisting of a microscopic analysis in the magnification range 100X–600X using cross-polarized light to evidence structural defects of the fibers like microcracks; and
(c) a preliminary mechanical test consisting of a microscopic analysis of the behavior of the single flax fibers subjected to bending to check macroscopic anomalies when one end is already glued on the tab.[29]

We do underline that this preliminary selection of the samples was necessary to avoid testing fibers with evident defects and/or contamination visible by means of an optical microscope.

---

[29]See footnote 14 on p. 213.

Table 7.2 List of flax samples used in the mechanical dating method not included in Table 6.1–Table 6.3

Picture	Name	Description	Dating	Origin	Collection	Notes
	NI	Roman age fabric 1–1 Mummy	220– 100 B.C.	Lima Peru	M. Moroni	[14]C dated
	NII	Roman age fabric 1–1	350– 230 B.C.	Engedi Cat. 133685- 120	Israel Antiquities Authority	[14]C dated

There were 11 series that had provided fibers capable of passing the above-mentioned selection procedure and precisely:[30]

- Two samples of modern flax fabrics, A and B, representing the present. As noticed above, they were used to perform a mechanical comparison for evaluating the bleaching effects with sodium hydroxide.
- Two samples, DII and D, representing the medieval age.
- the sample FII representing the Roman period of the first century A.D.
- Three samples, NI, NII (Table 7.2), and E, covering the first Roman age.
- Three samples, HII, K, and LII, representing the period ranging from about 1000 B.C. to 3500 B.C.

However the possible microcontamination, not visible under an optical microscope, can be investigated by evaluating the information derived from vibrational spectroscopy analysis (Chapter 6). From this point of view, we mention the case of the NI fabric (Table 7.2). The abnormal behavior observed in the vibrational spectroscopy tests[31] was also confirmed from the mechanical point

---

[30]If there is any specification following the series' name, the relative information is reported from Table 6.1 to Table 6.3.

[31]The sample does not appear among those described from Table 6.1 to Table 6.3, because it has been discarded due to the presence of contamination.

of view when processing the obtained data. In fact the measured mechanical properties turned out to be abnormal, too. After the analysis of both the mechanical data and those coming from the vibrational tests, we discarded the sample from the dating diagrams.

We can deduce that the mechanical dating method is beneficial when used in combination with other dating methods alternative to ^{14}C, such as the vibrational spectroscopy ones described in Chapter 6, and of course the other way around.

7.4 The Tensile Test

Tensile testing on single natural fibers is a widely used method in evaluating fundamental properties of textiles [79]. The tensile test is a destructive analysis carried out with a so-called tensile machine (Fig. 7.1) and with the aim of measuring the mechanical properties of materials in general.

The measured values could be reported in a force–displacement diagram, but a different graphic plotting is normally preferred. To make the tests independent of the geometric dimensions of the specimens (Section 7.2), the increment in length of the sample

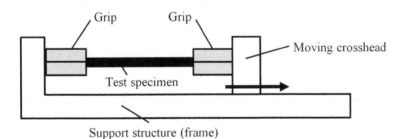

Figure 7.1 Schematic drawing of a tensile machine. The specimen is grasped to the device by means of grips in rest condition (zero load). A moving crosshead lengthens the test specimen and this produces a force perpendicular to its cross section. Both the force and the displacement are recorded during the test, thanks to proper measurement gauges. The test ends at the breaking of the specimen. A rigid structure (frame), characterized by negligible deformations compared to those shown by the specimen, supports all the devices of the tensile machine.

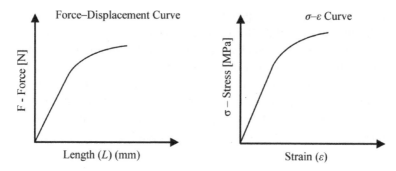

Figure 7.2   On the left qualitative example of a force–displacement curve, and on the right qualitative example of a stress–strain curve (σ–ϵ curve).

is divided by its initial length, thus obtaining the so-called strain (symbol ϵ). The force is instead divided by the initial cross section of the specimen and this physical quantity is called stress (symbol σ). Thus, the diagram is known as stress–strain curve or more simply σ–ϵ curve (Fig. 7.2). For obtaining the stress–strain curve, it is necessary to measure the initial length of the specimen and its cross section. We will see later (Section 7.7) how we determined them.

If, during the tensile test, we stop the increase of the force before breaking the specimen and, after unloading the sample, we perform several loading–unloading phases, these so-called cycles are useful for evaluating the damping of the material through the calculation of the loss factor (Section 7.2).

In addition to a rigid frame, the typical components of a tensile machine are in general:

- a displacement generator with a measurement gauge (the moving crosshead in Fig. 7.1);
- a gripping system of the specimen (the two grips in Fig. 7.1); and/or
- a force generator with a measurement gauge as, for example, a load cell (Section 7.1).

The ASTM D76-99 standard [13] defines both the three possible types of tensile test machines (purchasable on the world market or self-constructible) and the exact meaning of the various technical terms.

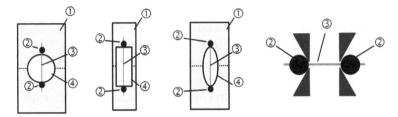

Figure 7.3 Examples of tabs adopted by researchers [10, 49, 86] following the reference standard [14] ASTM D3822-07. On a cardboard (1) a hole is made (4). The single fiber (3) is fixed with drops of glue (2). The dashed lines mark the cutting points when the tab is mounted on the grips of the tensile machine. The function of the two lateral bridges is to provide rigidity to the whole system during both handling and fixing the tab on the tensile machine. These bridges are cut away before performing the test. On the right, the original microdrops technique [149].

ASTM D3822-07 [14] is the reference standard for testing on single textile fibers. Among the tensile tests' guidelines in this standard there are those concerning the so-called tabbing technique, namely the manner of fixing the sample (namely a single flax fiber) onto a proper intermediate support (tab), which has a suitably wide area at its ends for letting the whole system be grabbed by the grips of the tensile machine.

The consulted literature shows that the reference standard has been widely followed. Figure 7.3 shows some adopted tabs types. Even our self-handmade tab (Fig. 7.13) partially derives from the models shown in Fig. 7.3.

Using a tab is not the only possible solution indeed. An original one is the so-called microdrops technique [155], adopted, for example, for testing single, short fibers such as those of wood and bamboo.[32]

For the machines, researchers, in large part, used two kinds of devices.

---

[32] Short fibers have lengths around 1.0 mm (0.039 in.). In the case of bamboo, some researchers [149] have deposited two drops of epoxy glue with a diameter of about 0.2 mm (0.0079 in.) at the ends of each fiber. The grips of the tensile machine (shown schematically with triangles in Fig. 7.3 on the right) were properly designed so as to use the drops as constraint points of the specimen and trough which transmitting the tensile force to the fiber.

On the one hand, a commercial tensile machine equipped with a load cell (the instrument for measuring the force acting on the fiber) is used with the lowest resolution[33] among those available.

On the another hand, an optical microscope and/or a scanning electron microscope coupled with a digital camera to measure the fiber's diameter is used by processing the obtained images with a dedicated software.

But there are also mechanical analyses on single fibers [149, 155] conducted by using a handmade tensile machine[34] built by the researchers themselves.

7.5 Measurement Challenges and the Adopted Solutions

An important measurement problem, already highlighted by the tests performed by the Max Planck Institute in 2009, concerns the commercial tensile machines' load cell used for carrying out mechanical studies on single fibers.

In fact we calculated that a load cell with a maximum capacity of 2 N (0.45 lbf) would be already used for a minimum part of its operating range in the case of fibers having a gauge length around a few millimeters (less than 0.1 in.) and diameters within 0.005 mm (0.0002 in.) and 0.025 mm (0.0010 in.), such as those of the Shroud and others derived from ancient fabrics.

Thus, taking advantage of structural and dimensional information and of the mechanical characteristics of flax fibers, we decided to achieve the following goals:

(1) Building a tensile machine at virtually no cost capable of providing the parameters for the stress–strain curve
(2) Measuring:

   - the breaking strength σ_R;
   - the strain at break ϵ_R;

---

[33] See footnote 9 on p. 211.

[34] The handmade tensile device used for investigating the mechanical behavior of single bamboo fibers has been constructed by purchasing the above-mentioned typical components. A computer processes the data and controls the machine's components.

- elastoplastic information on the fiber;
- sliding effects during loading;
- loading–unloading cycles in order to evaluate the loss factor; and
- the Young's moduli[35] E.

(3) Satisfying the following design characteristics:

- Measuring cycles in correspondence of stresses ranging within 0 and 2/3 of the breaking strength with resolution better than about 100 measurement points per cycle
- Measuring elongation of single fibers having serviceable lengths[36] from 1 mm to 30 mm (0.039–1.18 in.)
- Measuring elongation of 1% with a resolution of 0.0001 mm
- Measuring forces within 0 and 50 cN[37] (0–0.11 lbf), with a resolution of 0.0002 cN (0.00000045 lbf)[38]
- Measuring uncertainty of the measured results better than 10%,[39] which was eventually obtained

We computed the values assumed for the design specifications of the tensile machine by referring to a simplified model of a flax fiber with a constant circular cross section and elastic behavior characterized by certain features.[40]

---

[35]Since we wanted also to evaluate loading–unloading cycles, we considered other elastic moduli.

[36]See footnote 14 on p. 213.

[37]The notation means centinewton (cN), namely cents of newton (N). The utility of expressing the force in centinewtons instead of the most used newton unit in the SI system is that 1 cN is approximately equal to the weight produced by a mass of 1 g (0.035 oz.). Thus 50 cN corresponds roughly to the weight of a mass of 50 g (1.76 oz.) placed on the plate of a weigh balance. We will see later that in our tensile machine forces are measured by means of an analytical balance, which provides value in grams, that is, practically cN.

[38]This resolution is equal to 2 millionths of a newton. To have an idea of how slight is such a force, the reader considers that 0.0002 cN is roughly the weight of a 0.2 mg mass, which is more or less equivalent to that of a 2 mm (0.079 in.) long Shroud's flax thread.

[39]The design uncertainty is intended here as the target uncertainty that an engineer defines before starting his or her design and it is useful as a comparison among the various uncertainties encountered during experiment planning.

[40]Reference gauge length from 1 mm to 30 mm (0.039–1.18 in.); constant diameter ranging from 0.005 mm to 0.025 mm (0.0002–0.0010 in.); Young's modulus range of 1 to 100 GPa (145037–14503774 psi); and 1% design percentage strain at break.

A displacement resolution of 0.0001 mm (0.0000039 in.) requires a careful evaluation of the design solutions for the machine's elements that have to be built with tolerances,[41] deformations, and/or backlashes so small as not to cause undesired displacements on the specimen greater than the design resolution.

On the basis of the fiber's design reference data, we calculated both the maximum and minimum (i.e., the resolution) tensile design force that the machine must be able to measure: maximum force 50 cN (0.11 lbf); resolution 0.0002 cN (0.00000045 lbf). With such specifications the device must have a dynamic range of 250,000 points,[42] not easy to obtain from the technical point of view.

According to the design condition of virtually zero cost, we decided to start with the following simple solution. On the one side we built a handmade load cell by means of a strain gauge[43] (Fig. 7.4a) dynamometric cantilever beam. In fact all the required instruments were immediately available in our laboratory. On the other side we used a micrometer slide[44] (Fig. 7.4b) as a displacement generator.

We studied various solutions with different displacement generators. Figure 7.5 schematically shows the solution adopted in our tensile machine.

We faced the biggest measurement problems with the strain gauge cantilever beam load cell. Despite many calculations, we were not able to find feasible solutions with the equipment available in our Laboratory. Thus, searching for solutions with different load

---

[41] Tolerance is the geometrical or dimensional variability allowed by the project.

[42] It is obtained dividing the maximum force (50 cN) by the resolution (0.0002 cN).

[43] Strain gauges allow measuring of relative displacements between two generic superficial points of a loaded body. In a typical foil strain gauge a "constantan" metallic foil pattern (grid) is embedded in an insulating flexible backing. The deformation of the underlying surface of the body changes the electrical resistance of the grid. This resistance change is related to the strain that is usually measured using a proper electrical circuit and a power amplification of the signal.

[44] In this device the advancement or retraction of the micrometer screw's pin is obtained by turning the knob so that the pin moves in a sliding plane. The micrometer screw has a displacement range of 0–15 mm (0–0.59 in.) with a resolution of 0.010 mm (0.00039 in.). The micrometer has 50 divisions and a turn corresponds to a displacement of the pin of 0.5 mm (0.020 in.). Thus a one-notch rotation corresponds to a displacement of the pin of 0.010 mm (resolution).

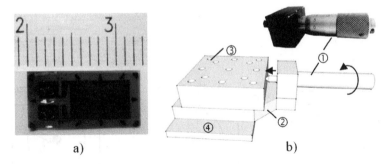

Figure 7.4 a) Picture of a strain gauge (the above reference scale is in millimeters); (b) schematic drawing of a slide micrometer: (1) micrometer screw, (2) base of the slide, (3) sliding plane of the slide moved by the micrometer screw, and (4) steel support plate.

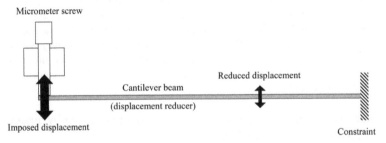

Figure 7.5 Schematic drawing showing the operating way of the cantilever displacement reducer. When translating, the pin of the micrometer screw bends the cantilever beam. In a cantilever beam the displacement of each cross section (a fraction of the imposed displacement) is related to that of the free end of the beam by means of a specific equation, namely each cross section moves downward by a quantity that is a function of its distance from the constraint. Thus, by choosing some suitable cross section, we can obtain particular reduction factors of the imposed displacement at the free end, such as, for example, 1/2, 1/5, 1/10, and 1/20.

cells, we decided to measure the force acting on a stretched fiber in terms of apparent loss of mass.[45] Figure 7.6 illustrates this idea in

---

[45] The weight is the force on an object due to gravity and it is equal to the product of mass for the local gravitational acceleration, which varies on the surface of the Earth. Whenever possible, in calculations the average value 9.80665 m/s^2 (32.174 ft./s^2), known as standard gravity, is used.

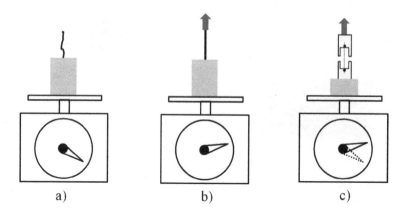

Figure 7.6 Example of force measurement through a single-plate weigh balance. A mass with an elastic attached on it is placed on the plate (a), and the balance measures its weight. Pulling up the elastic (b), the tensile force acting on it is opposite to the weight. Thus the balance shows an apparent loss of mass, which actually is a measure of the force stretching the elastic. If we replace the elastic with a single flax fiber (c), in the same way we can measure the tensile force acting on it as the difference between the mass of the system at rest and the gradually smaller mass values given by the instrument until failure.

a practical way: the experiment can be easily performed by anyone using a single-plate weigh balance.

We performed some tests in order to check the feasibility of this solution (the details are summarized in Section A.10 of the appendix). The results obtained from those tests were very good because:

(1) The tests conducted on both modern and ancient flax fibers have shown the feasibility not only of measuring the acting tensile forces but also of detecting loading–unloading cycles with 100 points.[46] Thus we were able to design and then to build a microcycling tensile machine.

(2) We have obtained know-how:

---

[46] As for the modern flax fibers (maximum measured force at break: 30.1 g, 0.0677 lbf) the first point of a cycle with 100 points would involve the measurement of forces around 0.1 cN (0.000225 lbf). In the case of ancient fibers the first point would require the measurement of forces around 0.001 cN (i.e., 1 mg, 0.00000225 lbf). All these mass values are detectable by a commercial analytical balances.

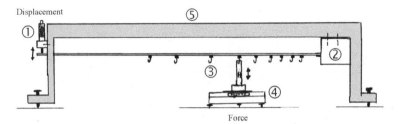

Displacement

Force

Figure 7.7 Design sketch [98] of the microcycling tensile machine: (1) displacement generator (micrometer screw), (2) displacement reducer (cantilever beam) equipped with hooks on which is hanging the specimen (a tabbed fiber), (3) tab with the additional mass, (4) force measurement gauge (analytical balance), and (5) frame (carrier beam).

- in extracting the fibers from the threads;
- in building a useful first model (Fig. A.12c) of our handmade tab; and
- in developing the tab's lateral-bridges breaking technique by means of a red-hot steel wire.

Figure 7.7 shows the original design sketch of the microcycling tensile machine. Its principal operating components are the following:

- *Displacement generator with measurement gauge*: It is designed with a cantilever beam (Fig. 7.5) flexed by the pin of the micrometer screw (Fig. 7.5) described above. The cantilever beam has nine hooks on which is hanging the tab. These hooks are glued in correspondence to the beam's cross sections in which the imposed displacement is reduced according to the nine appropriate reduction factors, including 0.05 (i.e., 1/20). The micrometer screw also performs the task of a displacement measurement gauge;[47]

---

[47] We chose a manual data acquisition, which now is automatized by means of a step motor micrometer screw and proper software.

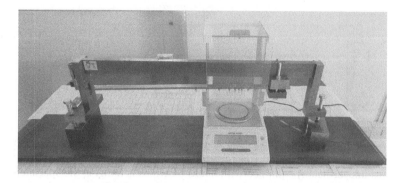

Figure 7.8 The realized microcycling tensile machine for single natural fibers. Depending on the reduction factor (namely the chosen hook), the analytical balance is shifted so that the vertical axis of its plate coincides with the axis of the hook to be used. The machine is horizontally controlled by means of two spirit levels, while the steel plate has the role of a rigid support for the whole system.

- *Force measurement gauge*: It consists of an analytical balance. The data were obtained manually by recording on a data sheet the value read on the display of the instrument;[48]
- *Fiber's tabbing system*: The lower end of a special tab is glued onto an additional mass to be designed in size according to the maximum capacity of the analytical scale.
- *Frame*: It is a portal structure horizontally regulated by screws. Both the micrometer screw and the cantilever beam are coupled with the frame.

7.6 The Microcycling Tensile Machine

The constructed microcycling tensile machine is shown in Fig. 7.8, while the drawing in Fig. 7.9 describes all its components.

---

[48] In many analytical balances it is possible to connect the instrument to a computer, but at the time we decided not to consider this solution. At present, we bought a different analytical balance; the mass values acquisition is totally automatized by using proper software.

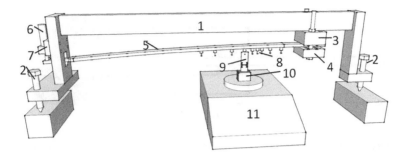

Figure 7.9   Sketch of the built microcycling machine: (1) frame horizontally regulated by screws (2); (3 & 4) clamping blocks for the displacement reducer cantilever beam (5); (6) micrometer screw gauge, mounted on a coupling plate (7), which moves the beam (5). To one of the hooks (8) a proper polyester mask (9) is hanged, tabbing the fiber under test. The lower part of the mask is glued with two iron blocks (10) that lie on the plate of the analytical balance (11).

We obtained the analytical balance[49] at "zero cost" (this was one of the project's goals). Since the analytical balance is an unmodifiable[50] component, we built all the other devices of the machine so as to adapt to the balance.

The fixed designed support of the displacement reducer cantilever beam (Fig. 7.7) provides a constraint theoretically preventing any movement of the fixed-end cross section of the beam. The adopted solution of building it with double resting blocks above and below the beam (points A and B) is shown in Fig. 7.10.

With the beam installed within the two resting blocks and the entire system connected to the portal frame, we glued the nine hooks (Fig. 7.11).

---

[49]Its specifications are resolution 0.1 mg (0.0000035 oz.) and maximum capacity 120 g (0.265 lb.). Thus this analytical balance has a dynamic range of 1,200,000 points, more than four times greater than the 250,000 design points. We thank Prof. Mirella Zancato for her advice and the Pharmacy Faculty of the University of Padua for allowing us to use the instrument.

[50]Actually the draft shield of the analytical balance might be removed and also its entire frame. But this would not be a correct choice. Once the draft shield is removed, the instrument would then operate without the structure protecting the plate by inevitable air movements (including those resulting from the breath of the experimenter), which can affect the measurements; in fact even a murmur over the plate is able to affect the measurements!

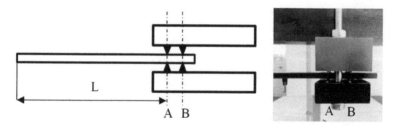

Figure 7.10 On the left, schematic drawing of the resting-block fixed constraint of a cantilever beam with free length *L*. On the right, realization of the constraint of the displacement reducer cantilever beam.

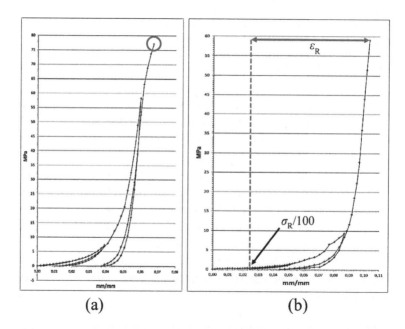

Figure 7.11 Particular of the hooks of the displacement reducer cantilever beam of the microcycling tensile machine; above is the corresponding nominal reduction factor.

The portal frame leans through regulating screws on two steel blocks that have the function of lifting the machine so that the distance between the analytical balance's plate and the hook is sufficient for hanging the specimen.

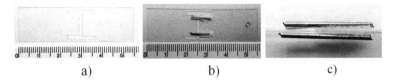

a)   b)   c)

Figure 7.12 (a) First tab model derived from a polyester sheet (Section A.10); (b) improved tab with misalignment-limiting devices made with iron bars. A paper strip is interposed between the iron bars and the mask in order to distance them from the tab's plane, but not too much as to render the system useless; (c) side view of the misalignment-limiting devices. The distance of the bars from the mask is about 0.5 mm (0.020 in.).

The actually used tab in the mechanical dating tensile tests is an improvement of the tab model realized (Fig. A.12c) during the feasibility tests of measuring forces by means of an analytical balance (Section A.10). The tab was characterized by a relative marked movement, above and below the tab's plane, of the horizontal edges of the handmade H-shaped carving. Such a movement often caused the breaking of the tabbed fiber. Thus, to reduce the possible planarity misalignments, we added four misalignment-limiting devices (Fig. 7.12b and Fig. 7.12c).

Figure 7.13 shows the manual construction of a tab.[51] Once a single flax fiber has been tabbed, the lower part of the tab is glued (Fig. 7.14) with two steel blocks (the so-called additional mass) that then lie on the plate of the analytical balance during the test.

From the engineering point of view the realized microcycling tensile machine is quite simple. Furthermore we achieved the design goal of building a tensile machine at virtually no cost, since its cost was less than 40 euros ($50)!

7.7 Data Processing

The values obtained from each tensile test (we briefly summarized the followed experimental procedure in Section A.12 of the appendix) must be processed for plotting the stress–strain curve

---

[51]The second author realized around 200 tabs.

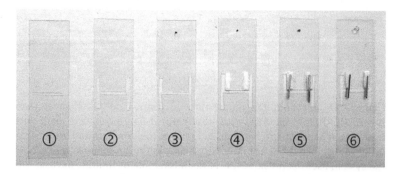

Figure 7.13 Phases of the polyester tab construction: dimensions 55.0 ± 0.3 mm × 16 ± 0.3 mm (2.165 ± 0.012 in × 0.630 ± 0.012 in.). (1) Rectangular cutting from a polyester sheet (form an A4 sheet 48 pieces can be obtained); (2) realization of H-shaped carving; (3) marking the position of the hole for hanging the tab to one hook of the displacement reducer; (4) gluing the paper strips of the misalignment limiting devices; (5) sticking the iron bars of the misalignment limiting devices; and (6) realization of the hole by means of a red-hot steel wire.

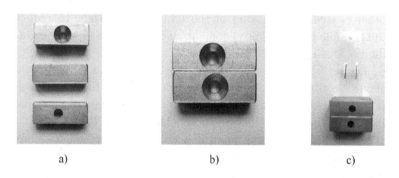

Figure 7.14 (a) Views of the steel blocks forming the so-called additional mass; (b) image of a double block obtained by joining two blocks with cyanoacrylate glue; (c) tab with the tabbed fiber equipped with the additional mass, which constrains the polyester mask's lower part on the analytical balance's plate during the tensile test. The total mass of two double blocks is about 77 g (0.170 lb.).

(Fig. 7.2) of the fiber and for measuring the desired mechanical properties (Section 7.5) from each σ–ϵ diagram.

To plot the σ–ϵ curve we need to know the fiber's initial length (gauge length) between the constraint points on the tab (glue drops)

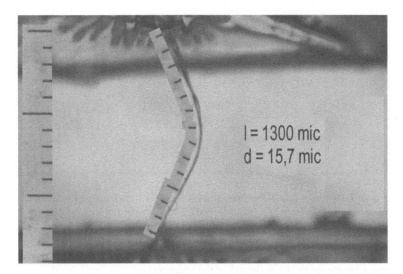

Figure 7.15 Technique used for determining the initial length of a single flax fiber (sample FII.13). We repeatedly reported on the photograph partial views of the millimeter scale associated with the series of photographs (electronic caliper), composing them in such a way as to follow the fiber.

and its diameter,[52] with which we can calculate the stress and the strain. We reached these goals by using an electronic caliper[53] with the aid of image processing software (Fig. 7.15 and Fig. 7.16).

Figure 7.17 shows the typical stress–strain diagram of a flax fiber obtained with the microcycling tensile machine. We immediately note that the shape of the σ–ϵ curve is in agreement with that reported in the literature, also in the presence of cycles[54] (Section 7.2).

---

[52] A flax fiber has a polygonal cross section (Section 5.1) with a central cavity (the lumen, Section A.11). Because of defects (Fig. 5.3), the breaking point is not always located in correspondence of the narrowest cross section. However, with a design uncertainty of the measured results not greater than 10% (see footnote 39 on p. 227), it is reasonable to consider the fiber as a solid cylinder with circular and constant cross section, thus represented by its "diameter." We measured the fiber's minimum diameter, following the simplifying assumption that the failure zone is located in this cross section.

[53] It consists of a vertical and a horizontal digital view of a millimeter grid (Fig. A.15b) with the same magnification factor used for photographing each lot of samples.

[54] In a tensile test a cycle is defined as any sequence of a loading phase followed by an unloading one (Fig. A.21a).

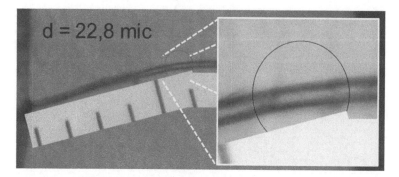

Figure 7.16 Example of the procedure followed for measuring the diameter of a single tabbed fiber (sample B.01). In correspondence of the cross section estimated with the smaller diameter we evaluated the pixel coordinates of the extremes of the segment digitally drawn on the fiber by means of image processing software. These data are then used to determine the diameter of the fiber (length of the segment).

The presence of loading–unloading cycles does not influence the general trend of the stress–strain curve.[55] This is a useful feature because it allows one to combine in a single tensile test also the study of the behavior of a fiber subjected to cyclic loading–unloading conditions.

The mechanical behavior of flax fibers is rather complex because the breaking strength depends on both the effective specimen length (gauge length) [105], the diameter of the fiber [36], and the absorbed moisture [139]. Thus we introduced a reference (or standard) flax fiber[56] and transformed (standardized) each tested fiber, different in size and in tensile behavior, as it would be the reference fiber. So, leaving unchanged the strain, we transformed the real obtained stresses into standardized stresses by multiplying them by three coefficients, respectively, evaluating the dependence

---

[55] In fact if we ideally connect with a segment all the peaks of the cycles at maximum load, the $\sigma-\epsilon$ curve would continue as if the tensile test was performed without cycles.

[56] The characteristics of this reference fiber are (1) length of the gauge length of 1 mm (0.039 in.), (2) diameter of 0.0150 mm (0.00059 in.), and (3) in contact and in equilibrium with air at a relative humidity of 50%.

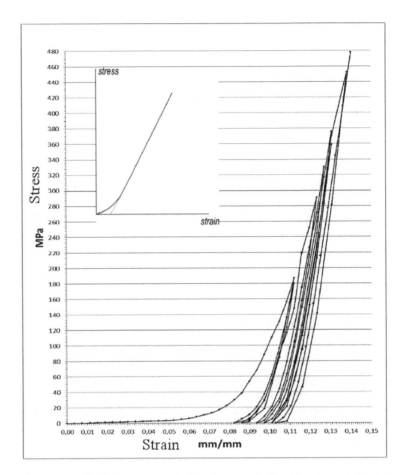

Figure 7.17 Typical stress–strain (σ–ϵ) curve of a flax fiber (sample DII.04) obtained with the microcycling tensile machine. We added, for a graphic comparison, a smaller image [8] with the qualitative shape of the σ–ϵ curve of a flax fiber, as it can be found in the literature.

on the gauge length, on the fiber's diameter, and on the moisture absorbed.[57]

---

[57] From the mathematical point of view we made an affine transformation. The real stress (σ^*) was standardized (σ) according to the equation

$$\sigma = K_L K_d K_H \sigma^*,$$

From the σ–ϵ standardized diagrams of the tested samples we evaluated the following mechanical parameters:[58]

- Breaking strength σ_R
- Elastoplastic strain at break ϵ_R
- Final elastic modulus E_f
- Elastic strain at break ϵ_E
- Decreasing elastic modulus E_c
- Direct loss factor η_D
- Inverse cycle loss factor η_I

In addition to the preliminary selection criteria (Section 7.3), to minimize the data scattering resulting from the presence of fibers' microdamage and/or contamination not visible in stereomicroscopic analysis, we introduced two exclusion criteria once the test had been concluded:

(a) Fibers with an evident anomalous stress–strain curve must be excluded.
(b) Each standardized breaking strength must be compared with the average one within each sample series. Whenever the breaking strength of a sample is outside the allowed variation range (between 2.5 times larger and 2.5 times smaller than the average breaking stress of the series), the sample must be excluded.

Table 7.3 summarizes the number of tensile-tested samples for each series and the final number of those used for identifying particular bijective dependencies with time.

---

where K_L, K_d, and K_H are the coefficients in the above-mentioned order and defined as follows:

$$K_L = \frac{1479}{1500 - 20.93L} \qquad K_d = \frac{67.62}{89.41 - 1.452d} \qquad K_H = \frac{743.0}{580.4 + 3.250H},$$

in which L (mm) and d (μm) are, respectively, the measured fiber gauge length and diameter and H)(%) is the laboratory humidity.

[58] For details on the definitions the reader can consult Section A.13 of the appendix.

Table 7.3 Number of samples subjected to a tensile test. The table shows only the flax fibers that passed the preliminary selection, rules (a), (b), and (c), discussed in Section 7.3. The causes of the tensile test failure were accidental breakage of the fiber during its gluing on the tab (tabbing technique), its positioning on the microcycling tensile machine, and cutting of the lateral bridges (Section A.12)

Name	Tabbed fibers	Not tested samples	Tested samples	Samples excluded by d)	Samples excluded by e)	Samples used for dating
A	21	7	14	2	0	12
B	19	4	15	2	0	13
DII	4	0	4	0	0	4
D	21	15	6	1	0	5
FII	19	9	10	0	0	10
NI	5	2	3	0	0	3
E	5	1	4	0	0	4
HII	5	2	3	1	0	2
K	8	5	3	0	0	3
LII	10	7	3	0	0	3

7.8 Results

Following the preliminary selection criteria of the samples (Section 7.3) and applying the two exclusion criteria to the data obtained from the tensile tests (Section 7.7), we evaluated for each series the average values of the breaking strength σ_R, the elastoplastic strain at break ϵ_R, the final elastic modulus E_f, the elastic strain at break ϵ_E, the decreasing elastic modulus E_c, the (direct) loss factor η_D, and the inverse cycle η_I loss factor.

Then, for each of the above-mentioned mechanical properties we placed the representative value of each of the nine tested series in a diagram with the years on the horizontal axis[59] in order to identify bijective correlations between the mechanical characteristics and the date in agreement with condition 2 (Section 6.1).

---

[59]What we did is qualitatively shown in Fig. 6.3 on the left. It is sufficient to imagine those diagrams with only nine points because nine are the tested series for each measured mechanical property Y.

The trend analysis allowed us to identify the following bijective property–time relationships:

- Breaking strength σ_R vs. time
- Final elastic modulus E_f vs. time
- Decreasing elastic modulus E_c vs. time
- Loss factor η_D vs. time
- Inverse cycle loss factor η_I vs. time

Supposing that natural cellulose degradation obeys [88] first-order reaction kinetics, namely an exponential decay law over time such as the one characterizing the radiocarbon decay (Chapter 4), we demonstrated that both the breaking strength σ_R and the two Young's moduli E_f and E_c follow an exponential decay law over time, too, determining the corresponding equation (condition 5) by means of the least squares fitting method.[60] In other words, these three mechanical properties exponentially decrease with the fabric's age. We of course considered other kinds of least squares interpolation curves, first of all the linear law and the polynomial one, but the exponential law was the curve that best interpolated the data according to the Pearson correlation coefficient.[61]

Taking, as we actually did, semilogarithmic coordinates in plotting the diagrams of the mechanical properties σ_R, E_f, and E_c, the above-mentioned exponential relationships with time turn out to be a straight line. We used this form in the diagrams for dating the Shroud. We point out that this transformation has no effect on the results, because it is simply a different way of presenting the data in a graph. For example, Fig. 7.18a shows the dating diagram of the Young's modulus E_f.

---

[60]We refer the reader to Section 6.1 for more details on this method.

[61]The interpolation curves with Pearson's correlation coefficient R are the following:

$$\sigma_R = 139.14e^{0.0009678x} \pm 336 \quad years \qquad R = 0.943$$
$$E_f = 6.2219e^{0.000709386x} \pm 418 \quad years \qquad R = 0.915$$
$$E_c = 8.1758e^{0.00060588x} \pm 537 \quad years \qquad R = 0.910$$

See footnote 12 on p. 193 for the meaning of the coefficient R.

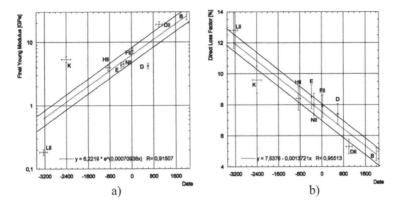

Figure 7.18 Two of the five dating diagrams [61] obtained by the authors: (a) dating diagram of the Young's modulus E_f; (b) dating diagram of the direct loss factor η_D.

Both direct loss factor η_D and inverse cycle loss factor η_I show a different bijective correlation with time, which is linear[62] instead of the exponential type. The loss factor (direct and inverse) grows with the antiquity of the sample. This would suggest that this mechanical property is not correlated to the resistance of the cellulose chains but rather to their degree of crystallinity. Since the degree of crystallinity is supposed to decrease with aging, a greater micromobility of the polymer chains of the fibers is allowed, with the macroscopic consequence of an increase of the internal dissipation energy when the fiber is subjected to cyclic loading–unloading conditions. Figure 7.18b shows the dating diagram of the direct (percentage) loss factor η_D.

We utilized the five dating diagrams with the relative uncertainty evaluation[63] (condition 7) for dating the Shroud as follows.

---

[62] The interpolation curves with Pearson's correlation coefficient R are the following:

$$\eta_{D\%} = 7.5376 - 0.0013721x \pm 193 \quad years \qquad R = 0.955$$
$$\eta_{I\%} = 4.2604 - 0.0011458x \pm 385 \quad years \qquad R = 0.900$$

[63] The lines above and below the central one shown in Fig. 7.18 delimit the uncertainty band.

Table 7.4 Estimates of the Shroud age provided by the five dating diagrams

Characteristic	Years
σ_R	577
E_f	14
E_c	456
$\eta_{D\%}$	−510
$\eta_{I\%}$	564

Fibers of the Shroud, coming from the back of the sheet (in correspondence of the gluteal area) and picked up during the 1978 Shroud of Turin Research Project (STURP) analyses (Section 2.3), after petrographic microscope identification by the first author (Section 5.3), have been tensile-tested by the second author.

Each dating diagram was used for evaluating a dating range of the relic. For example, once the value of the breaking strength σ_R of the Shroud on the regression line of the corresponding dating diagram (namely σ_R vs. time) was set, we calculated its abscissa, which is the dating of the sample. From the relative band we obtained the uncertainty. What we just described for the breaking strength has been done for each of the five mechanical properties in a bijective relationship with time (condition 9).

The obtained results are shown in Table 7.4 and Fig. 7.19, which also shows the final result, namely the dating of the Shroud, with its relative uncertainty at a confidence level of 95%, according to the ISO-GUM-BPIM guide [26].

A useful parameter able to assess the reliability of the used mechanical method is the dispersion of the results derived from the five mechanical datings. The more the resulting values are different from each other, the more we can suppose that various environmental factors might have negatively affected the results, providing values affected by not negligible bias. For instance, in some deteriorated fibers the dispersion of the results turned out to be excessive, with differences between the dates resulting from the same sample to be even around 2000 years.

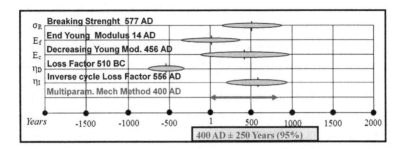

Figure 7.19 Dates of the Shroud resulting from the analyses of the five different mechanical parameters and combined result corresponding to A.D. 400 ± 400 years at a confidence level of 95%.

For the Shroud all the dates, except the one given by the direct loss factor, are compatible[64] with each other. This means that we obtained results containing dates around A.D. 400.

The differing result (represented by the dating from the direct loss factor) can be explained as follows. Both the slippages between the microfibrils and the plasticity of the fiber, not stabilized by suitable pre-tensions that have not always been possible during the tests, might have affected the result. Therefore in the case of the Shroud the relatively low dispersion of the results indicates that the final date is little affected by environmental bias and thus reliable.

We thought that the simple arithmetic mean of the five dating results, which gives for the Shroud the value A.D. 220, would be a less correct procedure. To combine into a single value the five values coming from the dating diagrams, we preferred a relationship[65] in which a greater weight is assigned to the mechanical properties less susceptible to plastic slippages. Hence the final dating value of the relic is evaluated as a weighted mean between the five results obtained from the dating diagrams.

---

[64]See footnote 10 on p. 138.

[65]The assumed equation is

$$Y = \frac{2Y_{\sigma_R} + Y_{E_f} + 3Y_{E_c} + Y_{\eta_{D\%}} + 3Y_{\eta_{I\%}}}{10}$$

in which the chosen weights favor the contribution of E_c with respect to E_f and of η_I compared to η_D. As regard the breaking strength's weight coefficient we assumed the value 2, intermediate between the minimum and maximum chosen values.

We observe that if we had used a weighted mean with Pearson's coefficients as weights, we would not have given greater weight to the mechanical parameters less dependent on plasticity, which is instead the criterion assumed to define the way of combining the five obtained dates together.

For the Shroud the MMPDM gives the date of A.D. 372. The computation of the uncertainty, associated with the value corresponding to the combination of the five dates, gives a value of ±400 years at a confidence level of 95%. Therefore, the dating of the relic is A.D. 372 ± 400 years (95%), but it can be rounded to A.D. 400 ± 400 years (95%). Hence the final result is compatible with the era in which Jesus of Nazareth lived in Palestine.

We also observe that this result is compatible (Section 9.1) with those provided by the chemical dating methods described in Chapter 6 and by the numismatic analysis described in Chapter 3, but it is not compatible with the 1988 radiocarbon dating result discussed in Chapter 4.

Part III

Something More about the Shroud

Chapter 8

Shroud Samples Spread for Scientific Research

An analysis by Marco Conca[1]

Since 1988, the year of radiocarbon dating (Chapter 4), no more samples have been officially taken from the Shroud, except from those in 2002 that, still, are in the exclusive custody of the archdiocese of Turin.

However, several scholars took advantage of the samples taken in the past years, and still they are doing it. This is the reason why it can be useful for the reader to know what these samples are on the basis of which the current research is carried out, waiting for new samples to be given to the scientific community.

More than focusing the attention on the path of the Sheet, here we will concentrate on the samples taken in the recent years, sometimes considered as relics, insofar part of the Sheet

---

[1] Marco Conca is a lawyer with a grounded experience in academic teaching developed in Switzerland, Italy, and also the United States. He is a great, passionate scholar of history of the Catholic religion. Meeting the first author in 2008 has been for him a turning point for focusing especially on the Shroud of Turin. Today he is an expert on its history.

The Shroud of Turin: First Century after Christ!
Giulio Fanti and Pierandrea Malfi
Copyright © 2015 Pan Stanford Publishing Pte. Ltd.
ISBN 978-981-4669-12-2 (Hardcover), 978-981-4669-13-9 (eBook)
www.panstanford.com

that enveloped Christ's body, sometimes as fragments of scientific relevance.

8.1 The Relics: An Overview

A relic is for definition the physical remains of a saint or his or her personal effects preserved for purposes of veneration. We can distinguish three different kinds of relics. First-class relics are related to items associated with Christ's life or physical remains of a saint. Second-class relics are instead something a saint wore, owned, or usually used in his or her life. Third-class relics are all the objects touched to a first- or second-class relic.

The importance of the Shroud stays in the fact that it not only was touched by Jesus Christ, but it also contains also a part of Him: His blood.

As far as the Shroud is concerned, the first fragments that appeared over the course of history with relic purposes date back to 1700. These relics are conserved in metal and glass shrines, sometimes closed with a cord, with sealing wax displaying the crest of the authority in charge of the closing operations.

The preciousness of these fragments, at that time, was clearly driven by faith, so much so that a need for producing new relics directly from the relics (creating the above mentioned third-class relics) spread.

Christianity relics are, now and in this particular case, very important tools for science. Faith is notoriously a gift, but it happens that who receives it, or who is waiting for it, could find in science a solid basis on which to build his or her own faith.

8.2 Finding Shroud Relics

Since 1988 the papal custodian of the Shroud, in agreement with the Vatican and the International Center of Sindonology of Turin, officially forbade new sample taking from the Shroud for scientific purposes.

Figure 8.1 Shroud relic of Orareo, considered probably authentic by the scholars Raymond Rogers and Giulio Fanti.

However, those samples are not the only ones circulating; in fact, we already know that the first fragments are dated back to 1700 and they have been used for relic function.

As a recent work shows, the American scholar Richard Orareo, for example, found a binder containing several eighteen-century relics (Fig. 8.1) and one of them was related to the Shroud.

He asked for other scholars' opinions; also Professor Fanti was able to analyze it, finding that it was compatible with some Shroud samples of proven origin in his possession.

Another sample taking comes from the restoration made by Sebastian Valfrè on June 26, 1694. On that occasion he sewed and mended some holes formed on the precious Sheet during the centuries. Obviously, some nonrecoverable threads remained in his hands.

Noteworthy is also the Shroud fragment shown in Fig. 8.2, owned by the Savoy family.

Who knows how many other chances the Shroud guardians had for "handling" it. What is certain is the thread taking, also in the area of the feet bloodstains, made from Professor P. Baima Bollone, in 1973, the year in which M. Frei took dusts and fibers from Shroud samples using adhesive tapes placed in contact with the Holy fabric.

Figure 8.2 Savoy's reliquary containing a little piece of the fabric (the dark part) taken from the Shroud.

The University of Genoa (Northwestern Italy) received some of those blood-taking samples in order to carry out the first studies on the Shroud DNA.

And here we were in 1978, when the American chemist Raymond Rogers was allowed, by the archdioceses of Turin, to take some material from the Shroud surface through particular tapes, together with G. Riggi di Numana, who drew the Shroud dusts, and M. Frei, who took other material always by means of tapes.

Afterward, another famous American chemist, Alan Adler, received from Rogers some of the above-mentioned tapes. The investigations on this material, together with the studies of Professor Bollone, showed, for the first time unequivocally, the presence of human blood particles on the Shroud!

Furthermore, Rogers was generous in distributing the extracted material; therefore, part of it was delivered to the Los Alamos Institute in the USA. Chemical-physical analyses on this fragment, then exposed at the Columbus congress in 2008, showed that the thread was half linen and half cotton. This result would have constituted a clear demonstration of the hypothesis that a mediaeval mend was sewn in the corner. The same corner from which the little strip that the radiocarbon dating test dated back to a historical time not compatible with the life period of Christ in Palestine was taken.

The United States was not the only country for Shroud research. The French engineer Marcel Alonso, well-known sindonologist, also received from Rogers one or more threads that he used for his studies related to the explanation of the Shroud image formation.

Also the French doctor Thibault Heimburger had further threads from Rogers.

Professor Fanti, as well, received from Rogers some linen fibers extracted through the mylar tapes in 1978, from the area of the body image, and some dozens of linen fibers coming from a thread taken by L. Gonella in 1978 (Fig. 8.3).

When Rogers died, his wife, Joan, gave all his material to the photographer Barrie Schwortz, who funded on January 21, 2010, the Shroud of Turin Education and Research Association, Inc. (STERA) and assigned to it the material received by Rogers.

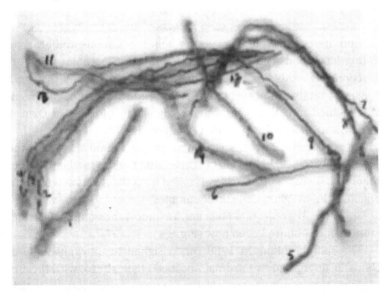

Figure 8.3 Little fragments recovered by L. Gonella after the sample taking in 1988.

Figure 8.4 Residual of the 1988 sample taking now exhibited in the Arizona State Museum.

And here we were in 1988, when the famous radiocarbon dating happened. The strips taken for the test were cut in halves (Section 4.2): the first, denominated "Riserva" (Fig. 4.3), was a fragment retained for eventual further studies; the other segment was divided into three parts and assigned to the labs of Oxford, Zurich, and Tucson for the radiocarbon dating.

As a matter of fact it is right and proper to add that since the piece of fabric given to the lab in Tucson was slightly smaller than the other two pieces, a little strip was taken from the Riserva in order to equalize the weights of the three samples.

So, in the Arizona State Museum, an important sample of Shroud fabric can be admired even now (Fig. 8.4).

The sample taking in 1988 was assigned to Giovanni Riggi di Numana, appointed by Cardinal Anastasio Ballestrero, with the aim of giving the samples to the labs for the radiocarbon dating. On that occasion, the Italian technician was also charged with the task of using a special vacuum extractor, already used in 1978 for collecting

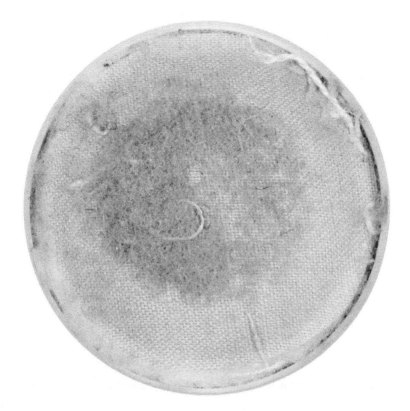

Figure 8.5 Filter "i," diameter 38 mm (1.50 in.), with the remaining extracted from the Shroud.

the dusts from the side of the Shroud sewn together with the Holland cloth (Fig. 8.5).

Also he had to take some little blood crusts in correspondence to the image of the nape. During his sample taking some residuals of the little cut linen strip were given to him.

Riggi di Numana had several meetings with Fanti. The professor decided to delve more deeply into the Shroud studies also encouraged by the Cardinal Giovanni Saldarini's letter written on May 26, 1997 (Fig. 8.6).

Torino, 26 maggio 1997

Prot. 691/96
Cat. 81

Chiarissimo Professor Giulio Fanti,

in merito alla Sua gentile lettera del dicembre scorso vorrei ringraziarLa per il Suo impegno di interessamento e ricerca a riguardo del problema Sindonologico.

Sono stato favorevolmente impressionato per le proposte che Lei mi sottopone e le ho trasmesse ai miei consulenti. Spero che ad esse si potrà dare seguito o nel 1998 o nel 2000. Forse Le potrà essere utile prendere contatto con il Centro Internazionale di Sindonologia (Via S. Domenico 25 - 10122 Torino), anche per un eventuale intervento al Congresso di Sindonologia che si terrà al termine della prossima Ostensione (5-7 giugno 1998).

Colgo l'occasione per assicurarLe il mio orante ricordo.

+ Giovanni Card. Saldarini

✠ Giovanni Card. Saldarini
Arcivescovo di Torino

Figure 8.6   Letter from Cardinal Saldarini, custodian of the Shroud, to Fanti.

Turin, May 26, 1998.

> *Dear Professor Giulio Fanti,*
> *With reference to your kind letter of past December I would like to thank you for your commitment in the interest and research as far as the Shroud issue is concerned.*
> *I have been positively impressed for your proposals and I have forwarded them to my consultants. I hope that they can have a positive follow-up in 1998 or in 2000.*
> *Maybe you can find useful to get contacts with the International Center of Sindonology (via S. Domenico 25, 10122 Turin) also for agreeing the possibility of your intervention during the Congress of Sindonology scheduled for the days after 1998 Shroud Public Display (5–7 June 1998).*
> *Ensuring my prayers for you.*
> *Cardinal Giovanni Saldarini, Archbishop of Turin.*

During one of their meetings, Riggi di Numana reported to Professor Fanti of having provided for free all the equipment needed for the sample taking, and also for this reason, the cardinal gave to him some Shroud samples, heartily recommending to *administrate that material as a good family man manages the assets for his children.*

The precious material received had not to be wasted; however, after the first phase of careful verification of Fanti's scientific work, trusting in him, Riggi di Numana willingly, gratuitously, and overall nonreturnably entrusted to him a little part of the material that in the following years would have led to the interesting scientific results published in this book.

The vacuum extractor was also lent to another researcher, the French engineer Alonso. After being ensured by Riggi di Numana that this extractor had not been used for other activies after 1978 and 1988 sample taking, the French engineer decided to pick up the particles in the inside part and in the plastic pipe that linked it to the various intake probes. From this process, Alonso succeeded in extracting a remarkable number of particles, including humans rests (little skin and blood crusts), organics (pollens, mites, molds), and minerals.

A very little piece of control was then given to Fanti, whereas Professor Gerard Lucotte, from École d'Anthropologie of Paris, could analyze a great number of particles and publish the results of his research on some scientific reviews. Moreover, he identified several human red blood cells, and also pollen and mineral particles typical of the area of Palestine!

On May 25, 2003, Riggi di Numana officially consigned the Shroud sample he collected in 1978 and 1988 to Fondazione 3M (T/N 3M Foundation) in which he founded the Center of Scientific Documentation on the Shroud (Fig. 8.7).

An extract of the Italian text in Fig. 8.7:

> The Center of technical and scientific documentation on the Shroud of Turin, financed by Fondazione 3M, placed in one of the institutional offices of the 3M Italia Spa is actually oriented to conserve the scientific documentation produced from the research carried out between 1978 and 2000 over the burial Cloth that tradition attributes to Jesus Christ.

Figure 8.7 Web page describing the center of technical-scientific documentation on the Turin Shroud, Chambéry, Lirey, just created by Riggi di Numana at Fondazione 3M di Milano, Segrate.

> *The systematic archive of this material has laic documentary function, therefore it is open to the audience. [...] The center contains more than 3000 paper documents, among which scientific publications, books, letters, articles, reviews, writings, authorizations and notes, beyond some samples and remains of the work executed between 1978 and 2000. There are also more than 5,000 photographs, someone unreleased, from reportages, scientific documentations and some clips in VHS and DVD that describe the research of scholars and multidisciplinary groups between 1978 and 2000 and the samples taking for the radiocarbon dating of 1988.*

By means of this center Fanti had the possibility to get bibliographic documentation first and some Shroud samples then.

The supply of sindonic material to Fanti by Riggi di Numana on behalf of Fondazione 3M started after that Fanti proposed to the

Fondazione 3M

16 MARZO 2004

Bolla di Consegna n. 2

Destinatario : Prof. Giulio Fanti del Dipartimento d' Ingegneria Meccanica
Università di Padova.

Indirizzo : Dipartimento di Ingegneria meccanica dell'Università di Padova – Via
Venezia 1 – 35131 Padova

Contatto : Tel 0409 827 6804   E mail – giulio.fanti@unipd.it

Descrizione del materiale :
n. 1 - Piccolissima quantità di polveri prelevate dal filtro "I" aderenti ad una striscia di
nastro adesivo sterile a bassa adesività inserite in n sacchetto di Pvc trasparente.
n.2 – Due piccoli frustoli ed un frammento di circa 2 mm. di lunghezza di filo di
Lino contenute nello stesso filtro "I". Questi campioni sono stati prelevati
singolarmente cercando di non contaminare il campione e senza aggiungere polveri
addizionate.

Nota bene - Questo materiale è stato consegnato a fondo perduto per effettuare prove
tecniche preliminari in relazione ad un progetto ancora embrionale, non
commissionato, pervenuto alla Fondazione 3M il 9 marzo 2004. In conseguenza a
questa bolla non seguirà alcuna fatturazione. Il destinatario è comunque pregato, e si
dichiara disponibile, a consegnare una relazione tecnica sulle prove effettuate alla
Fondazione non appena le prove saranno considerate concluse.

Per la Fondazione   Firma del Destinatario

Figure 8.8   First delivery note to Fanti of sindonic material from Fondazione
3M, March 16, 2004.

foundation, on October 20, 2003, a zero-cost study entitled "Analysis
of Microparticles Vacuumed from the Turin Shroud," as coordinator
of a research group comprising seven scholars.

So, on March 16, 2004, it was the first time (Fig. 8.8) when
nonreturnable sindonic material was given to Fanti with a document
produced by the Shroud Studies Group of the University of Padua
dated March 9, 2004, entitled "Proposal of an Analysis of Fibrils and
Little Fragments Coming from the Sample Takings Carried Out on
the Shroud of Turin."

Text of the delivery note to Fanti of sindonic material from
Fondazione 3M, March 16, 2004:

March 16, 2004
> *Delivery note no. 2*
> *Addressee: Professor Giulio Fanti, Department of Mechanical Engineering, University of Padua*
> *Address: Department of Mechanical Engineering, University of Padua, via Venezia 1, 35131 Padua*

> *Material description:*
> *n.1—Very little quantity of dusts collected from Filter "I" stuck to an adhesive tape sterile strip in transparent PVC bag.*
> *n.2—Two little fragments and one 2 mm length linen thread contained in the same filter "I". These samples have been extracted singularly trying not to contaminate the sample and without adding combined dusts.*

> *Please note, this material has been considered as non-returnable and it has been consigned in order to carry out preliminary technical tests in relation to a still embryonic project, not commissioned, arrived at Fondazione 3M on March 9, 2004. By consequence, this delivery note will not be invoiced. The addressee is asked, and declares himself available, to consign a technical relation to the Fondazione 3M on the tests carried out as soon as these tests are concluded.*

Fanti, as described in Delivery Note n. 2 of Fondazione 3M, on March 16, 2004, received the first samples consisting of:

- *very little quantity of dusts collected from filter "I" [...]*
- *two little fragments and one 2 mm (0.079 in.) length linen thread contained in the same filter "I"*

Please note, this material has been considered as non-returnable [...]
A dense correspondence followed.

For example, Riggi di Numana wrote in a letter to Fanti on March 30, 2004:

> *[...] I conclude with the recommendation of not divulging the re-start of the investigations: every word misunderstood or that could annoy someone could stop this new try. It is never good to sell the bearskin before having kill it.*
> *I wrote you a paper letter because this new adventure has to leave a mark for posterity. I think it is appropriate to suggest you to*

do the same if you consider that inserting a certain documentation in the Fondazione 3M archive about what you are doing in the Shroud research can be good and right.

Riggi di Numana asked Professor Fanti for confidentiality. The latter, respecting his will, preferred not to divulge the preliminary analysis results immediately.

As agreed in Delivery Note n. 2, Fanti consigned, on December 13, 2005, a 14-page document entitled "Preliminary Analysis and Comparison of Dusts Coming from Extracts Collected from Different Areas of the TS."

On July 12, 2006, Riggi di Numana, for Fondazione 3M, delivered to Fanti further material described in the delivery note with the same date. (Fig. 8.9).

Text of the delivery note for Fanti of sindonic material from Riggi di Numana, Fondazione 3M, July 12, 2006:

Following your request made on June 26, 2006, accompanied by a research proposal on the dusts coming from the rear of the Shroud in our archives, we deliver to you n. 5 pieces of adhesive tape on which we have, in your presence, stuck some micro-quantities of those dusts, in order to allow you and your lab to carry out the research.

The dusts consigned come from the filters used by G. Riggi di Numana in 1978 during the exploration of the Shroud rear and specifically from:

e—Hands area
f—Face area
g—Feet area
h—Buttocks area

To these, a micro-quantity of the indiscriminate marginal sample taking carried out with the same tools in 1988 has been added.

i—indiscriminate marginal sample taking of 1988

These samples are non-returnable and free, with the purpose of research and documentation. Nothing is due to G. Riggi di Numana or to Fondazione 3M for the material given, a part from the indication of its origin and source in the scientific publications that could derive.

Fondazione 3M

Spett Ing Giulio Fanti
Dipartimento d'Ingegneria Meccanica
dell'Università di Padova
Via Venezia 1
33131 Padova - I   Segrate, 12 luglio 2006

In seguito alla richiesta da Lei effettuata in data 26 giugno 2006, corredata da una
proposta di ricerca sulle polveri provenienti dal retro della S. Sindone ancora esistenti
nel nostro archivio, Le consegnamo n. 5 spezzoni d nastro adesivo su cui abbiamo, in
sua presenza, fatto aderire delle microquantità di quelle stesse polveri, al fine di
permettere a Lei e al Suo laboratorio di effettuare la ricerca progettata.

Le polveri consegnate provengono dai filtri utilizzati da G. Riggi di Numana nel 1978
durante l'esplorazione del retro sindonico e specificatamente da :

    e – Area delle Mani
    f – Area del Volto
    g – Area dei Piedi
    h – Area Glutea

a cui è stata aggiunta una microquantità del prelievo indiscriminato marginale
effettuato con gli stessi mezzi nel 1988

    i – Prelievo indiscriminato marginale del 1988

La consegna è effettuata fondo perduto, senza contropartite, per la ricerca e la
documentazione successiva (non è quindi richiesta la resa in qualsiasi forma) e nulla è
dovuto a Riggi o alla Fondazione 3M per il materiale concesso, salva la citazione
sulla sua origine e provenienza nelle pubblicazioni scientifiche che potrebbero
derivarne.

Distinti saluti

Firma per ricevuta :

Figure 8.9 Delivery note for Fanti of sindonic material from Riggi di Numana, Fondazione 3M, July 12, 2006.

Best regards
Giovanni Riggi di Numana

On August 10, 2006, Fanti sent to Fondazione 3M a 17-page detailed report entitled "Preliminary Analysis of Dusts and Little Fragments Deriving from Shroud Sample Takings Carried Out in 1988."

The interest for those results was so great that Fanti, on August 20, 2006, was able to produce a further 44-page report entitled "Statistics Analysis of Dusts Coming from Extracts Collected from Different Areas of the TS."

The report included such interesting information that it was introduced to the International Congress on the Shroud in 2008, Columbus, Ohio.

Unfortunately on January 6, 2008, Riggi di Numana passed away, after a heart attack that struck him when he was waiting to talk, after many years, with the Papal Custodian Archbishop Severino Poletto.

A few days before, Fanti had a conversation with his friend and again Riggi di Numana underlined how much he awaited that meeting in order to clarify the issue about the partial restitution of the sindonic material of Fondazione 3M.

From then on Fanti made several contacts with the above-mentioned foundation through its lawyer, Antonio Pinna Berchet. However, the lawyer, before going on, consulted the foundation council in order to define if and with which modalities to continue the tight scientific relationship founded during the years between Riggi di Numana and Fanti.

The latter, in fact, showing a huge amount of documentation, asked to carry on the interesting research undertaken, also for giving an alternative dating to the Shroud. Obviously, as suggested by the lamented Riggi di Numana, for confidentiality reasons, nothing written resulted in reference with this last idea of dating the Shroud; therefore the research and the agreements about it with the lawyer Berchet remained exclusively verbal.

After the council expressed to Berchet a favorable opinion in regard with going on with studies, he arranged a meeting in which also Fanti took part.

The final decision, taken also thanks to Father Gianfranco Berbenni who participated and sustained Fanti's project, was that the professor could have the access to the foundation archive, in date to be agreed, for following the Shroud material takings needed for carrying on the analysis (Fig. 8.10).

The legal representative Berchet, however, not being originally aware of the verbal agreement between Fanti and Riggi di Numana and after having consigned further material on April 11, 2008

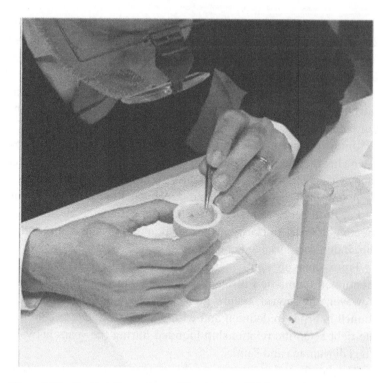

Figure 8.10 Fanti at the archive of Fondazione 3M during the extraction of some dusts from the filter H, vacuumed in the gluteal area on the back of the Shroud.

(Fig. 8.11), in order for Fanti to proceed with the established research, promptly interrupted any relationship, insofar he realized that these verbal agreements did not allow a direct control on the investigation by Fondazione 3M.

An extract of the delivery note of the sindonic material, April 11, 2008:

> *Dear Lawyer Pinna Berchet,*
>
> *I confirm our meeting scheduled on Friday, 11 April. I ask whether it is possible to better analyze the sample F15001 and another Shroud thread in addition to the samples of filters "i," "h" and "f" already agreed.*
>
> *Best regards*
> *Giulio Fanti*

Giulio Fanti
<giulio.fanti@unipd.it>
07/04/2008 18.47

To at.pinna.fondazione3m@mmm.com
     vittimia.fondazione3m@mmm.com
Cc
Bcc
Subject Incontro

Egr. Avv. Pinna Berchet,
confermo il nostro incontro fissato per venerdì 11/4 verso le ore 11.00 e,
con l'occasione, in riferimento alla relazione datata 23/1/2008
(Conclusioni: "Dall'analisi del campione di filo F15001 è risultato che
questo presenta una probabile contaminazione esterna consistente in
fibrille di cotone simili a quelle riconosciute da G. Raes come *Gossypium
herbaceum* ; sarebbe opportuno analizzare altri fili sindonici dell'Archivio 3M per
chiarire se effettivamente le fibrille di cotone possano essere solamente un
contaminante esterno come supposto dall'autore, in contrasto con quanto
rilevato in precedenza da G. Raes.. "), chiedo se è possibile analizzare meglio
il campione F15001 ed eventualmente un altro filo sindonico oltre ai
campioni dei filtri "I", "h" ed "f" già concordati.
Cordiali saluti.
Giulio Fanti

GIULIO FANTI Associate Professor
Department of Mechanical Engineering
University of Padua, Via Venezia 1, 35131 Padua - Italy
Telephone: +39-049-827-6804
Fax: +39-049-827-6785   /-0798
http://www.din.unipd.it/fanti

*Il giorno 11 Aprile 2008 sono stati
consegnati al Prof. Fanti :*

A — l'intero campione F15001

B — un campione del filtro I

C — un campione del filtro H

D — un campione del filtro F

*Dopo gli esami concordati , campioni
verranno restituiti alla Fondazione 3M*

Pinna Berchet   Prof. Giulio Fanti

Figure 8.11 Delivery note of the sindonic material, April 11, 2008.

On April 11, 2008, Professor Fanti received:
A—the whole sample F15001
B—a sample of filter I
C—a sample of filter H
D—a sample of filter F
After the tests, samples shall be restituted to the Fondazione 3M.

As written in the letter by Riggi di Numana on March 30, 2004, he
preferred not to put into writing preliminary agreements that time
could have helped to manifest in a more appropriate way.

After the 2008 delivery, the lawyer Berchet personally informed Professor Fanti that he could keep for himself all the material received, handling the studies in first person and by consequence without the support of Fondazione 3M, and that he never should mention anymore the results eventually reached.

On January 18, 2008, also the son and successor of Riggi di Numana wrote a letter to Fanti, declaring:

> The Department of Mechanical Engineer of the University of Padua and Professor Fanti are not authorized to cite the name of my father Giovanni Riggi di Numana in their scientific publications, unless after having obtained the official approval of the Fondazione 3M and his scientific committee; Riggi family does not ask the restitution of material of which professor Giulio Fanti refers in his letter of past 7 January.

This confirmed, therefore, as already the foundation did by means of the lawyer Berchet, that they did not demand the restitution of the sindonic material given to Fanti.

Fanti received the advice of his American friend and lawyer Mark Antonacci about the way in which he could behave toward the position of Fondazione 3M.

The American lawyer ensured the full property of Fanti on the Shroud material he had received, suggesting also to contact his colleague and friend, the lawyer Marco Conca, author of these pages, so that he could confirm all his conclusions on the basis of the Italian and Vatican law.

Antonacci asked and obtained two Shroud threads by Fanti in order to allow a common friend, the scientist A. Lind, to carry out further investigations of sample contamination, in particular intended to determine the percentage of rare isotopes of sodium and chlorine.

Meanwhile, Alonso, after the meeting with Fanti at Fondazione 3M on March 16, 2004, on behalf of Archeothonia, a company he founded, brought the vacuum extractor used in 1978 and 1988 to France to extract all the residuals of drawn dusts. A material and results exchange started between Archeothonia and the University of Padua.

It is at this point that the already mentioned Professor Gerard Lucotte entered the scene. He did not have any positive reply to his request of material to Fondazione 3M, so he directly asked Alonso in order to have some vacuumed dust samples. Since Alonso did not have any Shroud fiber yet, he replied to ask Fanti. The latter initially gave two or three Shroud fibers, but then, on the basis of Riggi di Numana's suggestion, interrupted any relation and did not provide any other material.

The peregrinations of Shroud fragments intertwine; run far and wide among universities and laboratories, scientists, heirs, and acquaintances; and are of interest to foundations, religious organizations, and enthusiastic people. The Saint relics are an object of vulgar commerce on the web and objects of mystification by the side of people who would like to have them but cannot. The more an object is rare and precious, the more his relocation is bound by mystery and tension.

Professor Fanti is a scientist. And as a scientist, the source of research has been objective, with a certain traceability of the studied relics. This is the reason why he minutely conserved every document proving the origin of the samples received.

The precision that always has distinguished his work and his research, the meticulous filing of data, evidence, and documents, is the objective tool that gives certainty of moral conduct, scientific validity of the results, and perfect synthesis of faith and science.

8.3 The Vatican Issue: From the Relics to the Shroud

Religion and science seem to be two worlds apart. Science is based on evidence; religion, instead, is based on faith, and faith is notoriously a belief not based on proofs.

The human race increasingly needs to believe; better if it can believe that what faith cannot explain can be explained by science. The great questions of life always intrigue all of us, especially now that we live in a present in which the word "progress" rules. Both scientists and the Catholic Church are aware of it and in the last decades the two realms are closer than ever. There is been a great openness of the Church to the new scientific discoveries and many scientists believe that God created the world through science.

The relationship between science and religion is complicated but not impossible. In general they travel along two parallel planes that intertwine at a certain time by comparing their results and, as in the case of the Shroud, show full compatibility.

As far as the relics are concerned, the debate is important both for scientists and even more for the Catholic Church.

This is the reason why it is noteworthy to dedicate a little space to the official communications and to the rules that the Catholic Church issued over this topic.

First of all, the Code of Canon Law, as far as the proper use of relics is concerned, decrees:

TITLE IV: THE VENERATION OF THE SAINTS, SACRED IMAGES, AND RELICS

Can. 1188—The practice of displaying sacred images in churches for the reverence of the faithful is to remain in effect. Nevertheless, they are to be exhibited in moderate number and in suitable order so that the Christian people are not confused nor occasion given for inappropriate devotion.

Can. 1189—If they are in need of repair, precious images, that is, those distinguished by age, art, or veneration, which are exhibited in churches or oratories for the reverence of the faithful are never to be restored without the written permission of the ordinary; he is to consult experts before he grants permission.

Can. 1190—§1. It is absolutely forbidden to sell sacred relics.

§2. Relics of great significance and other relics honored with great reverence by the people cannot be alienated validly in any manner or transferred permanently without the permission of the Apostolic See.

§3. The prescript of §2 is valid also for images which are honored in some church with great reverence by the people.

Monsignor M. Frisina, Rome Curacy Liturgical Minister director, stated in an interview granted to Romasette.it:[2]

---

[2] It is the online version of the weekly published by the Dioceses of Rome.

Figure 8.12 One example of a relic on sale on the eBay site: threads declared to have come from the Shroud, "De Sindone Dni. Nstri Jesu Christi," sold for $1000.

> *A Christened body is a body-temple of the Holy Spirit, but that of a saint is even more because he lived his sanctity in his flesh, a communion of grace with God, and his body has been pervaded from the same grace solemnly. The relic allows us to keep a sort of contact with the body's saint. During history, relics had an important role also in the fight against the Evil spirit, because the relic, being a physical reality that had a special relationship with grace, is not loved by devil.*
>
> *…It is absolutely forbidden to sell or buy any kind of relic because they are a sacred thing, therefore they do not have any price. The problem of relic selling is very spread in the Web and I must say that this is a desecration.*

Actually, despite these clear regulations, it is quite easy to find in the online market various Shroud relics on sale for some thousands of euros, as displayed in Fig. 8.12, and also threads taken in the past centuries. More astonishing is that among the sellers there is a person who seems to allude to have bought a lot of relics and sacred objects direct from the Vatican! This "father_salvatore" (his nickname) seems to have already sold 3026 objects on the eBbay site till October 2013. That represented in Fig. 8.13, a nail allegedly part of the Christ's Passion sold for $2800, is an example.

Vatican reliquary 1700s relic Holy Nail passion Jesus Christ COA

Authentic official Vatican real first class relic of the Holy Nail from the True Cross of our Lord Jesus Christ. The relic is a nail, a copy made in 1700s with some shavings from the real genuine Holy Nail of our Lord found at Holy Sepulchre and exposed at the Basilica Holy Cross in Jerusalem, Rome, with the original wax seal that assures the integrity of the relic. The relic comes in the antique wooden and crystal reliquary to expose the Holy Nail, leather outside and silk inside. Reliquary size approx 11x2.5x3.5 inches, 28x6.5x9 cm, rare and unique item available only to Vatican, lifetime guaranteed, with a certificate of authenticity signed by monsignor Nivardo Fiorucci, Prior Abbot of the Basilica Holy Cross in Jerusalem. On the base of the reliquary there is a sticker with the number of a Vatican private collection inventory. eBay policy prohibits the sale of human remains and requires a disclosure of what the relics are: these relics are iron, which are allowed by eBay policy. They are sacred and devotional relics of the Church.

Figure 8.13 One example of a relic on sale on the eBay site: a nail declared to have come from the Passion sold for $2800.

Just for curiosity, on October 29, 2012, Fanti sent the following email to the telematic seller; obviously, till today, no reply has come:

> *Hello, according to the Code of Canon Law: "Can. 1190 - §1. It is absolutely forbidden to sell sacred relics," my question is if I can buy this relic without committing a sacrilege. Thank you in advance for your kind reply. Regards.*

As far as Shroud samples are concerned, the official custodians of the Shroud made two official statements about the scholars' possession of samples. This custodians are Cardinal Saldarini and Archbishop Severino Poletto.

Cardinal Saldarini, in September 1995, made the following statement:

EXPERIMENTS AND ANALYSIS CONCERNING THE HOLY SHROUD

Turin, September 1995

News concerning research conducted with various purposes on supposed Shroud material has lately appeared in mass media. The Papal Custodian of the Shroud feels that it is his duty to express his position on this. More and more reports are being published about experiments carried out on Shroud material aiming to check the results of the tests conducted with the Carbon 14 method in the summer of 1988. This purpose may be legitimate, and the Church recognizes to every scientist the right to carry out appropriate research in their field of science. However, given the circumstances, it is necessary to point out that:

(a) no new sample of material has been taken from the Holy Shroud since 21st April 1988, and according to the Holy See and the Custodian of the Holy Shroud it is highly unlikely that a third party may be in possession of any residual material from that sample;

(b) if such material exists, the Custodian reminds everybody that the Holy See has not given its permission to anybody to keep it and use it and he therefore demands to give it back to the Holy See;

(c) it is not possible to know with any degree of certainty whether or not such experiments were carried out on material coming from the fabric of the Shroud. Therefore, the Holy See and the Papal Custodian declare that no serious value can be recognized to the results of those alleged experiments;

(d) this is evidently not the case of the research carried out on material taken for the tests in October 1978 and following the explicit authorization of the Papal Custodian;

(e) within the climate of mutual trust established with scientists, the Holy See and the Cardinal Archbishop of Turin encourage scientists to be patient until a clear and systematically planned research program is arranged.

Fourteen years later, on May 4, 2009, Cardinal Poletto, archbishop of Turin and papal custodian of the Shroud, resubmitted the same declaration of Cardinal Saldarini, though with some changes:

EXPERIMENTS AND ANALYSIS CONCERNING THE HOLY SHROUD

Turin, 4th May 2009

News concerning research conducted with various purposes on supposed Shroud material has lately appeared in mass media. The Papal Custodian of the Shroud feels that it is his duty to express his position on this, following the guidelines of the statements already made by his predecessor, Cardinal Saldarini in September 1995.

The Church does not dispute the fact that scientists have the right to conduct appropriate research on the Shroud, as long as they are respectful of people's faith and sensitivity and the rights of the Holy See, as the owner of the Shroud. However, it is necessary to point out some aspects about the use of such supposed Shroud material:

- No new samples of material have been taken from the Shroud for research purposes since 21 April 1988. According to both the Holy See and the Custodian of the Holy Shroud, it is unlikely that a third party may be in possession of any residual material from that sample. What was obtained during the restoration work of 2002 was immediately inventoried and put under seal, for exclusive use of the Holy See.
- It is not possible to know with any degree of certainty whether or not such experiments were carried out on material coming from the fabric of the Shroud. Therefore, the Holy See and the Papal Custodian declare that no serious value can be recognized to the results of those alleged experiments.
- Within the climate of mutual trust established with scientists, the Holy See and the Archbishop of Turin encourage scientists to be patient until a clear and systematically planned research programme is arranged. As far as this is concerned, the research programme is to be considered on hold until the end of the forthcoming exposition. It is to be hoped that it will start to be planned in the period of time which will follow.

It is not easy to understand why Archbishop Poletto forgot point (d) of Cardinal Saldarini's statement, in which he points out that the Holy See and the papal custodian cannot recognize *"serious value to the results of those alleged experiments,"* but *"this is evidently not the case of the research carried out on material taken for the tests in October 1978."*

It seems that while Cardinal Saldarini rightly recognized the scientific validity of the Shroud material taken during the 1978 analyses, Archbishop Poletto forgot the existence of this material, notwithstanding the official authorization of this sample taking.

In regard to point (b), unfortunately, the demand of Cardinal Saldarini to give the alleged samples back to the Holy See seems to have not the expected result. In fact, many scientists all over the world are still in possession of the samples received.

Only two are exceptions, one is that put forward by the undersigned, on Fanti's stead. We will broadly treat this issue afterward.

The second exception is related to fabric samples taken and given for studies to the textile expert G. Raes during the analyses carried out by the Pellegrino Commission in 1969–1973. In fact, Raes gave officially the samples back to the papal custody.

Despite the two cardinals' statements, apparently quite restrictive, it is noteworthy that Pope Emeritus Benedict XVI and the Holy See were interested in the new studies and research on the Shroud.

The proof of that comes from the official reply by Mons. Peter B. Wells, assessor of Vatican City, dated March 31, 2012, to the letter from Fanti, dated March 20, 2010, who wrote to the Pope:

> [...] *The University of Padua financed 54,000 Euro for a research project entitled: "Multidisciplinary analysis applied to the Turin Shroud: study of the body image, of possible ambient pollution and of micro-particles capable to characterize the linen fabric" of which I am the coordinator. The research group is composed of 5 professors and 3 researchers by Padua University and 4 professors of other academic department and other staff.*
>
> *The multidisciplinary study covers different fields and is divided into different points, related to the formation mechanism of the body image with formulation of the most reliable hypothesis; the analysis of the possible causes of pollution of the TS samples taken in 1988, for radiocarbon dating purposes; the analysis of the mechanical and morphological features of the linen fibers contained in the TS; the analyses of microparticles present in the TS.*
>
> *Given the importance of this project concerning the most important Relic of Christianity, the present is for asking a meeting with His Holiness, during which I would like to show in more details the studies that I mean to carry out and His opinion would be very appreciated. In the same occasion I also hope to have the time to illustrate the interesting scientific news discovered by my search group in the last years.*

SEGRETERIA DI STATO

PRIMA SEZIONE · AFFARI GENERALI

Dal Vaticano, 31 marzo 2010

Egregio Signore,

con cortese lettera del 20 marzo corrente, Ella ha indirizzato al Santo Padre espressioni di ossequio e devozione e, nell'informare circa un progetto di ricerca di codesto Ateneo, ha avanzato una particolare richiesta.

Nel ringraziare per il gentile gesto e per i sentimenti che l'hanno suggerito, il Santo Padre formula cordiali auspici di proficua attività culturale e professionale e, mentre affida Lei e quanti collaboreranno a questa iniziativa alla materna intercessione della Vergine Santa, di cuore imparte la Benedizione Apostolica, pegno di ogni desiderato bene, volentieri estendendola alle persone care.

Circa l'istanza, La invito a rivolgersi all'Em.mo Arcivescovo di Torino, Custode Pontificio della Sindone.

Con sensi di distinta stima

Peter B. Wells

Mons. Peter B. Wells

Assessore

Figure 8.14   Reply from the Vatican assessor to the letter of Fanti.

Waiting for a favorable reply from His Holiness, I pay my most regardful respect to Him.

In fact, the assessor, in the Holy Father's name, *expresses the best wishes for a fruitful professional and cultural activity* referring to the research project on Shroud studies (Fig. 8.14).

Text of the reply from the Vatican assessor to the letter of Fanti:

From Vatican State, March 31, 2010.

Dear Sir,

With your kind letter of March 20, you addressed to Holy Father expressions of reverence and devotion and, by informing Him about a university research project, you asked for a particular request.

The Holy Father thank you for the kind action and for the feelings that led to it and formulates benevolent wishes for a fruitful cultural and professional activity. While he entrusts you and all the people who are going to cooperate in this effort to the maternal intercession of the Holy Virgin, he blesses you by His heart with the Apostolic Benediction, extending it to your closest people.

About your request, I invite you to ask to the Archbishop of Turin, Papal Custodian of the Shroud.

With distinguished respect

Mons. Peter B. Wells, Vatican Assessor

8.4 The Restitution of the Shroud Sample in Possession of Giulio Fanti

Given the sincere friendship and the common interest in Shroud studies, Fanti assigned the undersigned to contact the Holy See and the Shroud Custodian in Turin, both in order to inform the authorities about the new scientific discoveries and in order to arrange the return of the Shroud sample in possession of the professor.

The archdioceses of Turin replied suggesting that we could return the material to Professor G. Ghiberti, president of the Diocesan Commission for the Shroud of Turin.

In agreement with Fanti, on August 20, 2012, I replied that, see Fig. 8.15, also because of a previous "misunderstanding" between Professor Fanti and Professor Ghiberti, *in order to avoid any further possible misinterpretation*, it would be preferred and considered more appropriate, *given its unquestionable importance, to return the Shroud material directly to the Papal Custodian of the Shroud.*

An extract of the fax from the lawyer Conca to the archdioceses of Turin:

Figure 8.15   Fax from the lawyer Conca to the archdioceses of Turin.

To Don Mauro Grosso
For Mons. Cesare Nosiglia
Archbishop of Turin

Following the letters of June 13 and July 4, 2002, supposing that there has really been some misunderstandings between Professor Fanti and Mons. Ghiberti, we prefer that this kind of "misunderstanding" will not be prolonged in the future meetings. So I consider useful to sum the last communications.

In the e-mail from Mons. Ghiberti, October 5, 2008 he insults without any reason Professor Fanti saying he lacks of scientific professionalism and moral uprightness. Professor Fanti does not deserve this. I deduce that mons. Ghiberti forgot point (d) of the letter from Cardinal Saldarini, published in September 1995 in which he explicitly declare that the research carried out with material taken in 1978 are considered official (www.shroud.it/FOSSATI.PDF).

In the e-mail from Mons. Ghiberti, October 24, 2008 he defines the clarifications that Fanti writes in the e-mail of October 10 as a "conversation between two deaf people".

Therefore, in order to avoid any further possible misinterpretation, I specify that the Shroud material can be returned.

Not receiving any reply, following the proposal of maintaining a "scientific dialogue" with the International Center of Sindonology, we reiterate the request of examining the high-resolution photographs taken by Haltadefinizione in 2008.

With best regards

Lawyer Marco Conca

Since then, we have been lying in wait for their reply ...

And here is that if we want to believe in "lucky" circumstances, just shortly before the printing of this volume, Turin authorities indirectly got in touch with Professor Fanti and, by informal and unofficial means, justified their silence, communicating a veiled disdain for having been contacted even by a lawyer. The restitution of Shroud samples should have been a simple and spontaneous action and overall *without any condition*. A standard mail delivery, practically!

Fanti asked for my mediation more for our friendship than that for my profession. The scientist, exactly because of the importance of these relics, felt the need for giving a correct, transparent, and precise formality to this handover.

It is extremely important, given the relevance of this matter, that no doubts be raised on who has what, on who is in possession of sure and guaranteed relics or objects of scientific investigation.

(We already talked about the paradox of finding online sellers on the eBay site that try to sell Shroud fragments!)

By the way, Fanti accepted to return the material, following the instructions given by Turin archdioceses.

Actually, a possible cooperation with the International Center of Sindonology in Turin would allow many scientists awaiting, since decades, to carry out scientific studies on the Shroud to satisfy their expectations.

Despite these practical difficulties that for the moment are preventing scientists from conducting further analyses, as already said, there are open windows.

In fact, on July 8, 2014, Assessor Mons. Wells wrote a letter to Fanti on behalf of Pope Benedict XVI (Fig. 8.16). The text declaims:

Dear Sir,

Pope Emeritus Benedict XVI [...], with the intention of showing His great satisfaction for the studies and research that you are

SEGRETERIA DI STATO
──
PRIMA SEZIONE · AFFARI GENERALI

Dal Vaticano, 8 luglio 2014

Egregio Signore,

il Pontefice emerito Benedetto XVI   ... omissis ...

Sua Santità, nel manifestare vivo compiacimento per gli studi e gli accertamenti che proseguono nel conoscere sempre meglio l'autenticità dell'importante Reliquia, mentre auspica che l'accurato studio in parola possa avere gli esiti sperati e dare un valido contributo alla ricerca scientifica e teologica, assicura il ricordo nella preghiera per le sue personali intenzioni e di cuore imparte la Sua Benedizione.

... omissis ...

Peter B. Wells

Mons. Peter B. Wells
Assessore

Egregio Signore
Prof. GIULIO FANTI
Dipartimento di Ingegneria Industriale
Via Venezia, 1

35131 **PADOVA**

Figure 8.16   Letter written to Fanti on July 8, 2014, from Assessor Mons. Wells on behalf of Pope Benedict XVI.

> *carrying out in order to recognize the authenticity of the important Relic, while he wishes that the accurate study you described will be able to have the hoped results and give a valid contribution to the scientific and theological research, ensure His prayers for your personal intentions and from His heart He blesses you. [...]*
> *Mons. Peter B. Wells, Assessor*

8.5 Historical Summary of the Shroud Samples Available Today throughout the World

For the reader more interested in the path of Shroud samples in the last decades, here is a synthesis that divides the samples into five periods: ancient samples and those taken in 1969–1973,

1978, 1988, and 2002. However, because of the obvious reasons of confidentiality, this summary could be not properly completed.

8.5.1 *Ancient Samples*

In the past, it was common to make relics also from pieces and threads extracted from the Shroud; therefore we can find several samples sealed in reliquaries dated back to 1500, 1600, and 1700. In addition to these samples there are also the third-class relics, obtained by touching the first- or second-class relic with a piece of fabric. In the text two examples are described, the reliquary of Orareo (Fig. 8.1) and the Savoy relic (Fig. 8.2). During the restoration carried out by Blessed S. Valfrè we can think that other relics have been produced.

8.5.2 *Samples Taken in 1969–1973*

During the analysis carried out by the Pellegrino Commission in 1969–1973, several samples, also blood samples, were extracted and analyzed by the following researchers: M. Frei, G. Raes, G. Frache, and G. Filogamo.

Raes gave the samples back, but it seems that the collections of adhesive tapes containing blood, fibers, and pollen was sold by Frei's wife after his death to the American researcher Alan Whanger.

8.5.3 *Samples Taken in 1978*

The samples collected in 1978 are probably the most interesting, and in agreement also with Cardinal Saldarini's letter, they maybe are the only ones that officially have been given to the researchers for study purposes.

This "gift" from the papal custody was given to the fact that the Shroud of Turin Research Project (STURP) researchers preferred not to sign a "straitjacket contract" proposed by the Turin curia, in which they asked that all the material acquired during the analysis be consigned to the papal custodian, forbidding them therefore to carry out the interesting analysis that later led to remarkable results published in several scientific specialized reviews.

These samples mainly consist of:

- adhesive tapes containing linen fibers, little blood crusts, and other material, all taken by different areas of the Shroud by the STURP scholar Rogers;
- vacuumed dusts from the space between the Shroud and the Holland cloth by the equipe of Riggi di Numana;
- little blood crusts taken by the equipe of Professor Bollone;
- ten or so threads taken by Professor Gonella; and
- adhesive tapes taken by Frei, similar to those sampled in 1973.

Some of these samples have been distributed for free to some scholars throughout the world. For example, Gonella gave his Shroud threads to Rogers, who sent them to T. Heimburger, J. Marino, and Alonso.

Rogers gave some of his tape samples to Alonso and Fanti, who in turn gave them to the various laboratories.

When Rogers died, most of the material was donated to the American group STERA founded by the STURP official photographer B. Schwortz.

Riggi di Numana, instead, gave, always for free, most of the material contained in the filters of the vacuumed dusts to Bollone and a little part to Fanti and Alonso; other material was consigned to L. Garza-Valdes. After Riggi di Numana's death, this material, officially in possession of Fondazione 3M, properly sealed by a notary, seems now conserved in the library of the University of Milan.

8.5.4 *Samples Taken in 1988*

During the extraction of the little strip of Shroud fabric aimed to the radiocarbon dating test, other samples were taken. Among them, were the dusts that Riggi di Numana was allowed to vacuum from the Shroud.

Not all the material addressed to the dating, however, was used; for example, a little piece of fabric is still conserved in the Arizona State Museum.

Another piece of the strip taken, called "Riserva," is conserved by the archdioceses of Turin for possible future controls on the results achieved in 1988.

8.5.5 *Samples Taken in 2002*

As already said, during the heavy intervention of "restoration" of the Shroud in 2002, the patches sewn by the Poor Clair nuns in 1534 have been removed, the areas burned by the famous fire have been scraped off, and some threads have been cut, especially from the back of the relic. All this material, properly sealed, is now conserved by the archdioceses of Turin, but till now it is not available to the scientific community.

Also some photographs and some spectra obtained from the secret analyses that have been carried out are not yet available to the scientists who required them.

Chapter 9

Recent and Future Developments

Even though the Vatican has not allowed scholars to take new physical samples of the Shroud since 1988, several scientists around the world continue to carry out research on the most important relic of Christianity, even basing their studies on the official samples taken in 1978. Some might say that these samples are obsolete, but since they are preserved in appropriate containers and isolated from the external environment, in many cases they are still very useful, especially for research at the microscopic level. Proof of this comes from members of the Shroud Science Group, which organized another conference on the Shroud in the U.S. in 2014, following the one held in Dallas in 2005.

This chapter brings together some of the more interesting recent discoveries and proposes new possibilities for future research. After summarizing the results regarding the alternative dating of the relic, the chapter deals with both the organic and the inorganic particles vacuumed from the Shroud.[1] This is followed by a presentation of results from analyses of bloodstains and some results from both plant and human DNA analyses.

---

[1] We refer the reader to the end of Chapter 2 for some notes on the foundation and the activity of this group of international and multidisciplinary studies on the Shroud led by the first author.

The Shroud of Turin: First Century after Christ!
Giulio Fanti and Pierandrea Malfi
Copyright © 2015 Pan Stanford Publishing Pte. Ltd.
ISBN 978-981-4669-12-2 (Hardcover), 978-981-4669-13-9 (eBook)
www.panstanford.com

9.1 New Dating of the Shroud

To confirm the unreliability of the radiocarbon dating carried out in 1988 and discussed in Chapter 4 several alternative dating methods have been developed.

Chapter 6 described the techniques of two chemical dating methods based on vibrational spectroscopy and presented the results. From a chemical point of view, using suitable diagrams (Figs. 6.9 and 6.11) vibrational spectroscopy makes it possible to identify molecular chemical bonds or groups of atoms characteristic of the material analyzed. Even polysaccharaides[2] in the flax fibers, first of all cellulose, as it is the main constituent (Chapter 5) in them, can be studied using these techniques.

The basic idea is that time degrades the fibers' polymers,[3] changing their chemical structure such that the concentration of certain groups of atoms, which are typical of cellulose, vary with the age of the sample. Vibrational spectroscopy can identify and "count" these groups.

After making a bias[4] correction of 452 years due to the Chambéry fire (Section 2.1), the dating of the Shroud using Fourier transform infrared (FT-IR) vibrational spectroscopy analysis turned out to be 300 B.C. ± 400 years at a confidence level of 95%.[5]

The dating of the relic from Raman vibrational analysis was 200 B.C. ± 500 years, again at a confidence level of 95%.

Both vibrational datings are compatible with the first century A.C., when Jesus of Nazareth lived in Palestine.

In Chapter 7 an alternative dating method was proposed, which was not chemical but rather mechanical; the description was a summary of the master's degree thesis in mechanical engineering [98] of the second author of this book under the supervision of the first author. Before developing this alternative method it was necessary to design and build a tensile machine for vegetable textile fibers (Figs. 7.9 and 7.10), like those of the flax plant, to be able to carry out mechanical tests on them.

---

[2] See footnote 7 on p. 168 for the meaning of this word.
[3] See footnote 6 on p. 168 for further information.
[4] This concept has been illustrated in Section 4.3.
[5] See Section 4.1 in which the meaning of confidence level is explained.

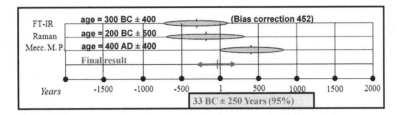

Figure 9.1 Date of the Shroud resulting from the analysis of three different methods, Raman, FT-IR, and mechanical, and the combined result corresponding to 33 B.C. ± 250 years at a confidence level of 95%. As described in Chapter 6, it should be noted that the FT-IR dating has been corrected to +452 years for the bias due to the Chambéry fire.

The basic idea in this case is that the degradation of the polymer chains of the fibers promoted by time, by breaking the chains and changing the order in which they are mutually arranged in space, can modify the mechanical properties in a such a way as to be able to carry out dating.

Indeed, the result was that five mechanical properties vary in a bijective[6] way over time. The multiparametric mechanical dating based on these five significant parameters combined together has led to the age of the relic of A.D. 400 ± 400 years with a confidence level of 95%.

At this point it is important to make a reasoned comparison of the results in order to understand if they are able to provide reliable information about the dating of the Shroud. The results reported in Fig. 9.1 are a summary of the combination of the mechanical results and those based on chemical spectrometry.

The mean of the values from the two chemical datings and the mechanical one indicate that the most likely date of the Shroud is 33 B.C ± 250 years with a confidence level of 95%.

It is important to note that the lowest uncertainty[7] assigned to the final result comes from directly applying the statistical rules recommended [26] by the current international norms. From a qualitative point of view, this can be explained by the fact that

---

[6]See footnote 9 on p. 187.
[7]See Section 4.1 in which this term is explained.

additional information coming from several dating values reduces the final uncertainty.

If the intervals of uncertainty of all three datings are considered, we finally observe from Fig. 9.1 that there is a common time interval, which is precisely the first century A.D., the time when Jesus Christ lived.

In addition to these three datings, it is important to note that the numismatic analysis reported in Chapter 3 demonstrates that the Shroud was used as a model for Byzantine coins (gold *solidi*) that were coined after 692. Therefore, the radiocarbon dating from 1988, which dates the Shroud at 1325 ± 65 years cannot be reliable, as the statistical analysis discussed in Chapter 4 has shown.

Four independent analyses (FT-IR, Raman, multiparametric mechanical, and numismatic) all make it possible to maintain with relative certainty that the Shroud is contemporaneous with the time of Jesus Christ. Therefore, it is possible that Jesus might have produced the double body image on the linen cloth, which still nowadays has neither been reproduced nor even explained.

9.2 Organic Particles

Some basic information about the so-called dusts that were vacuumed from the linen cloth of the Shroud was described in Chapter 5. Here the discussion moves beyond the flax fibers to consider various materials that are present in the dusts vacuumed from the relic.

During the analyses carried out by the Shroud of Turin Research Project (STURP) in 1978 and when samples were taken in 1988 for radiocarbon dating (Chapter 4), numerous areas were vacuumed in the interspace between the original linen cloth of the Shroud and the Holland cloth (present at the time but then removed in 2002), that is, in the areas that are not visible (Fig. 5.8).

The research group at the University of Padova, led by the first author together with Roberto Basso, Irene Calliari, and Caterina Canovaro, carried out new investigations on the vacuumed dusts. The group has refined the technique used to identify the fibers of the Shroud (Fig. 5.10) using cross-polarization analyses by means of a

Table 9.1 Classification of the material contained in the vacuumed dusts

Organic material	Inorganic material
Flax fiber resulting from the Shroud	Red particles (ferrous)
Cotton fibers	Whitish particles (calcareous)
Fibers of the Holland fabric	Dark particles
Other fibers, red and blue	
Pollen	
Spore	
Mites	
Red-brown particles similar to blood	
Particles of orange color	
Yellowish particles	
Plant and human DNA	

suitable petrographic microscope (Section 5.3). A significant amount of material that can be classified, as seen in Table 9.1 [55], was found using this technique.

To carry out a more detailed analysis of the vacuumed dust, the dust was placed on special adhesive tape on gold-plated aluminum disks called stubs and studied using an electronic microscope (Fig. 9.2). The resulting images were published at the Valencia Congress [57] and are shown in Fig. 9.3.

Some cotton fibers were found next to the flax fibers (Fig. 9.4), but the origin of these has yet to be investigated. Nonetheless, it is possible to imagine that at least some cotton fibers can come from the Shroud, since a few rare cotton fibers have been found mixed into the fibers of the Shroud.

Since the interesting palynological[8] analyses, carried out by the famous criminologist Max Frei, have been called into question and the scholar has even been accused of fraud, efforts were also made to find pollen in the dust vacuumed from the Shroud.

Unfortunately, the dust analyzed did not have as many pollen grains as the adhesive tapes studied by Frei, which came from the frontal face of the relic. However, some of the grains that were found can at least partially confirm the analyses carried out by the Swiss

---

[8]See footnote 1 on p. 143.

Figure 9.2 Example of a stub, an aluminum disc about 10 mm (0.394 in.) in diameter, containing dust drawn from the Shroud and metallized for subsequent analysis under an electron microscope.

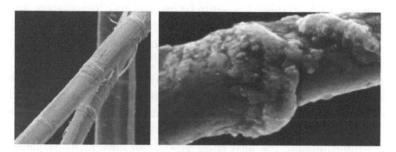

Figure 9.3 To the left, a modern flax fiber observed under the electron microscope, compared with a flax fiber of the Shroud, on the right, which appears encrusted. Spectrographic analysis has also shown that this encrustment has a calcareous nature.

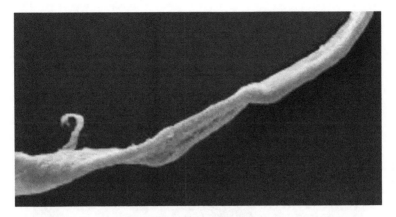

Figure 9.4 Cotton fiber with the typical ribbon-like shape found in the vacuumed dust coming from the Shroud.

scholar, who was unable to complete his research before his sudden death.

One of the main accusations from scholars who do not believe in the authenticity of the Shroud against Frei's research is fraud because the photographs of the pollen that he published in scientific articles do not refer to pollen from the Shroud but rather correspond to more recent pollen grains. Clearly if a researcher is found to be a fraud, all of the results of his analyses would have to be considered irrelevant.

A justification of this apparently correct accusation has already been given by the fact that at the time of publication it was more difficult to take photographs with an electronic microscope. Therefore, it was common for scholars to publish photographs of pollen grains that were as similar as possible to the samples being studied.

Proof of this explanation comes from the analysis of the vacuumed dusts on the stub shown in Fig. 9.2: a pollen grain of *Phyllirea angustifolia*, an evergreen plant that flowers between March and May and adapts well to the difficult terrain of some Mediterranean areas that are characterized by extreme drought. This type of pollen was just the type classified by Frei in his work.

Confirmation of this comes from the fact that the pollen grain seen in Fig. 9.5, which comes from the Shroud dust, has

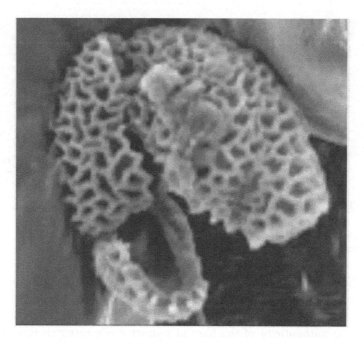

Figure 9.5 Pollen grain of *Phyllirea angustifolia* found in dust vacuumed from the Shroud.

clearly decayed over time, making it easy to think that the Swiss criminologist may have found other grains that were similarly decayed. This is why he preferred to publish integral images of current pollen grains rather than images of similarly decayed fragments of pollen. Obviously, even if the pollen grain shown in Fig. 9.5 is a pollen fragment, there are no doubts about what kind of grain it is. This debunks one of the main accusations of fraud brought against Frei, and therefore, the interesting analyses he carried out can be considered reliable.

Among the other pollen grains, a probable grain from an ash tree (Fig. 9.6) was found; this type of tree is typical in various parts of Europe. A pollen grain relative of the Lebanese cedar tree (Fig. 9.7) was also found. Although the cedar can be found in various areas around the Mediterranean, it is typical of the Middle Eastern areas and can, therefore, lead back to the areas where the Shroud was on display about 2000 years ago.

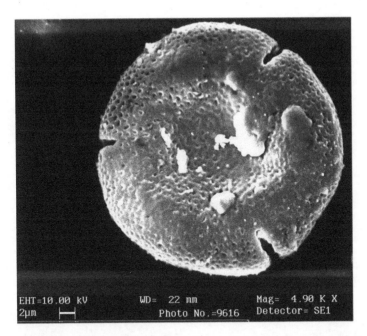

Figure 9.6 Probable pollen grain of *Fraxinus angustifolia* found among the vacuumed dust from the Shroud.

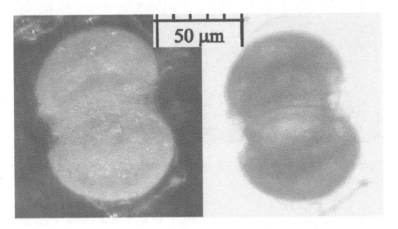

Figure 9.7 Pollen grains of *Cedrus libani* found in the vacuumed dust from the Shroud.

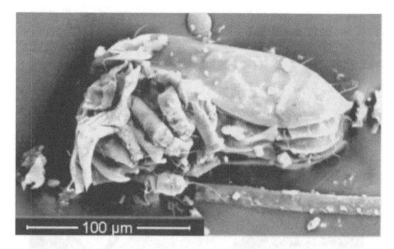

Figure 9.8 Example of a mite found among the vacuumed dust of the Shroud.

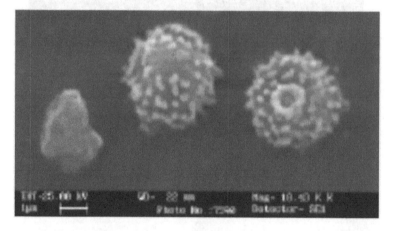

Figure 9.9 Example of spores of *Aspergillus glaucus* found in the vacuumed dust of the Shroud.

When mites (Fig. 9.8) and fungal spores and moss such as (Fig. 9.9) were found, it was clear that the Shroud needed special conservation efforts such as those that were carried out during the 1990s to eliminate the parasites that were damaging the linen cloth. As we have seen in Chapter 1, the Shroud is now kept in a controlled case in an atmosphere of inert gas with very little oxygen in order to avoid the proliferation of these parasites.

9.3 Inorganic Particles

It is not always easy to distinguish between organic and inorganic particles since at first sight they can appear to be fairly similar even if the crystalline shapes characterize the latter. Even for the first author of the book, for example, it was not initially easy to distinguish between bloodstains that appear in various shades from red-orange to dark red and mineral particles that contain iron oxide and are thus reddish.

Using a spectroscopic technique called energy dispersive X-ray spectroscopy (EDS) it was possible to distinguish between different types of particles by analyzing the atoms in the particles being examined. For example, with this technique, organic particles can be distinguished from other ones because of the greater presence of oxygen and carbon atoms.

From an analysis of the dust samples on the adhesive tapes taken from the area of the feet during the 1978 sample taking carried out by STURP, Joseph Kohlbeck found a higher concentration of calcium carbonate (containing traces of strontium and iron) compared to the other areas of the Shroud that were studied. After having analyzed samples from the soil in Jerusalem, which proved to be very similar to the dust studied, he concluded that the Man of the Shroud must have walked barefoot on the ground in Jerusalem before being buried. This conclusion, perhaps a little hasty, led to debate among scholars; therefore, the first author tried to verify, at least partially, what the American researcher had claimed to have proven.

Researchers at the University of Padova studied particles whose size ranged from 0.003 mm (0.000118 in.) to 0.30 mm (0.0118 in.). Many mineral particles were identified, but those worthy of note were higher quantities of minerals belonging to the *illite*[9] and

---

[9] Illite belongs to the clay-sized micaceous mineral, phyllosilicates subgroup.

Figure 9.10 On the left, mineral particles coming from Jerusalem: the lightest at the top come from excavations in the Holy Sepulchre; the darker ones below are reddish iron-rich particles from Mount Zion. On the right, particles vacuumed from the Shroud: the clearest are the ferrous one, with a size of the order of a thousandth of a millimeter (0.000039 in.).

smectite families (both clay minerals), clay particles, and a few rare quartz particles.

At the same time, a research study carried out in collaboration with the geologist Amir Sandler from the Geological Survey of Israel in Jerusalem showed that these minerals were typical of the soil in Jerusalem[10] and other Mediterranean areas subject to the desert winds of the Sahara.

It was interesting to compare (Fig. 9.10) the dusts vacuumed from the Shroud with those taken both from excavations carried out underneath the Holy Sepulchre and from the soil of Jerusalem (Mount Zion).

The first author was very much surprised to find significant similarities between the various particles analyzed. For example, as can be seen from the comparison of the spectra in Fig. 9.11, which show, respectively, a particle gathered at Mount Zion and a particle

---

[10]Sandler has identified particles of illite–smectite, containing large amounts of iron, which are typical of the clay of Jerusalem. These particles have been found both in the Shroud's dusts and in the samples of soil collected on Mount Zion. In addition to these particles, in the dusts coming from the Shroud Sandler has found particles of calcite and apatite, containing chalk, talc, quartz, potassium, dolomite, titanium, goethite, and hematite, which are all typical of the soil of Jerusalem.

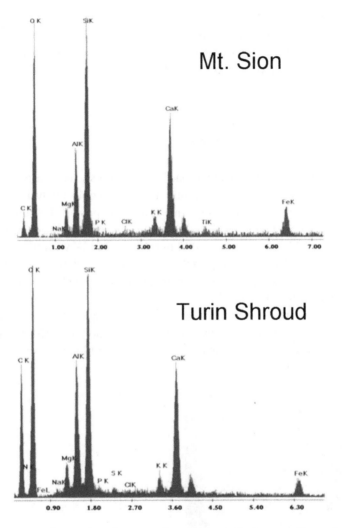

Figure 9.11 Comparison between the spectrum of a ferrous silicate from Mount Zion (above) with a morphologically similar particle coming from the Shroud's dusts (below): the two spectra match.

taken from the Shroud, it is clear that the material analyzed appears the same. On the basis of this result, we cannot and do not want to say that the dust from the Shroud comes without a doubt from Jerusalem, because there are other regions in the Mediterranean and the Middle East that contain similar minerals in the soil. However, once could also make the argument that the body of the Man of the Shroud probably fell to the ground repeatedly, picking up soil from Jerusalem that was then left on the Shroud when the corpse was enveloped in the cloth.

9.4 Little Crusts of Blood

As we have written, it is not easy to distinguish reddish inorganic particles from little crusts, but a spectroscopic analysis would be of great help to highlight the atoms composing the substance in question. Figure 9.12, for instance, shows a little crust of considerable size—about 0.030 mm (0.0012 in.)—with the related spectrum. The gold (Au) peak is due to the fact that before the observation through the electronic microscope, the big particle has been metallized with gold.

It is worth noticing that the spectrum of Fig. 9.12 is very similar to the spectra obtained by Pierluigi Baima Bollone [15], forensic scientist of Turin, who analyzed fragments of blood directly taken from the Shroud of Turin.

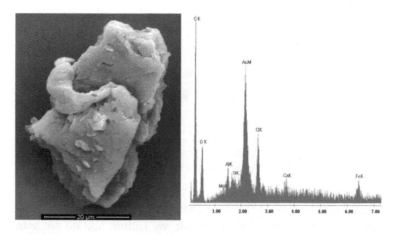

Figure 9.12 On the left, a large particle of blood viewed under the electron microscope and on the right its spectrum, typical of the blood.

According to the American researchers J. Heller and A. Adler [78], the samples of human blood of the Turin Shroud contain several elements: carbon (C), oxygen (O), sodium (Na), magnesium (Mg), aluminium (Al), silicon (Si), phosphorous (P), sulphur (S), chlorine (Cl), potassium (K), calcium (Ca), and iron (Fe).

The spectroscopic elemental analysis carried out on the red biological particles coming from the dusts vacuumed from the Shroud reveals the presence of C, O, Si, Mg, Cl, Ca, and Fe. All these elements there are in the blood and body fluids involved in the various enzymatic and metabolic processes.

For example, chlorine is particularly present in extracellular fluids, calcium is essential for the process of coagulation, and iron is contained in hemoglobin. On the basis of the percentages of these elements found in the blood of the Man of the Shroud, in-depth medical analyses that are to be made in the future will determine the physical conditions of the Man, also in relation to the tortures he underwent.

9.5 Vegetal DNA Analysis

Professor Gianni Barcaccia, expert geneticist of the University of Padua, was consulted by the first author with regard to a possible study on the dusts of the Shroud. What Professor Barcaccia suggested was a separate analysis of vegetal, animal, and human DNA, in this way sampling not a single specimen but a set of dusts taken from a specific area. According to the geneticist, if the Shroud of Turin truly is the sheet that was exposed for thousands of years in diverse environments and preserved following different procedures, the dusts it attracted had to be full of traces of DNA left by contact with various human beings.

To carry out this study several samples of dusts on the order of 0.01 mg (0.00000035 oz.) had been taken and independently treated with the so-called molecular baits able to isolate the specific DNA of a variety of human beings. The analysis consisted of highlighting particular sequences of the plastidial[11] and mitochondrial[12] genome[13] with the purpose of identifying the organisms where they came from.

With the exception of a bird extinct centuries ago that might have contaminated the Shroud when it was publicly exhibited in the past, Professor Barcaccia, with Giulio Galla and Michela Verna, had succeeded in isolating vegetal and human DNA, but not animal.

Therefore, even from the DNA point of view there is no presence of vegetal traces mixed with animal ones, as further proof of the fact that the fabric of the Relic contains only vegetal fibers (most of all linen mixed with cotton) in accordance with Jewish Law that forbids mixing vegetal and animal fibers.

As for the vegetal DNA [58], to identify the species of origin, the analysis had been made through the amplification of specific gene and intergene sequences typical of plastidial genome (chloroplasts) and nuclear genome (chromosome). The parts of the DNA were rapidly multiplied by using a technique named polymerase chain reaction and were studied by means of a technique of sequencing able to highlight the particular nucleotidic structure.[14] After this analysis, the main worldwide databases of genes and genomes had been addressed to identify the species of origin.

About 100 samples were amplified through this technique and it was possible to reconstruct the nucleotidic structure of about 60 of them, identifying 24 different plant species (the list is included in Section A.14 of the appendix). Many of the species detected, like clover (*Trifolium repens* and *Trifolium fragiferum*) and rye-grass (*Lolium multiflorum*), are widespread in the Mediterranean basin, including in the Palestine; others originate from Central Asia but spread up to the Middle East, like the pear tree (*Pyrus*) and the plum tree (*Prunus*). The latter two are related to the species *Pyrus*

---

[11] Plastids are a group of cellular organelles that characterize plant cells responsible for several activities connected to cellular metabolism.

[12] A mitochondrion is a little organ of the cell that generally has an elongated shape (reniform or with the shape of a bean). It is present in all the eukaryotes, that is to say, living unicellular or multicellular organisms made of cells with a nucleus.

[13] The term genome indicates the sum of the genetic materials of an organism.

[14] Nucleotides are simple molecules composed of three substances: a nitrogenous base, a five-carbon sugar, and a phosphate group. They are guanine, adenine, thymine (uracil in RNA), and cytosine.

cossonii, a tree native of Algeria, and *Pyrus syriaca*, a pear tree spread in Turkey and Syria.

Even more interesting for the reconstruction of the historic journey of the Shroud of Turin is the English plantain (*Plantago lanceolata*), a native of Palestine. There are then other species, spread over vaster areas (including Europe), which attest the exposition of the Shroud in different geographical areas.

In agreement with Professor Barcaccia, the importance of the DNA analysis lies in the great diversity of the species identified, which is atypical if compared with similar studies that always present a much reduced number of species. They can be related to Asian, African, and European areas. This is an indication that the Shroud was displayed in very different areas ranging from cultivated plains to mountainous areas, from arid to more myths environments.

9.6 Human DNA Analysis

As for the human DNA [58], no attempt was made in this initial phase of analyzing the DNA of the Man of the Shroud, but what was studied at first was the possible level of contamination of the Shroud due to the contact with the human body (saliva, tears, hands touching it) and the corresponding places of origin in terms of ethnic groups.

In the past, analyses on these sort of materials found on the Shroud had seldom been carried out. Perhaps the only scientific result that deserves attention dates back to 1995 [33], when a team led by Professor M. Canale of the University of Genoa did some research in collaboration with the above-mentioned Professor Bollone: crusts directly taken from the Shroud were studied. At the end what came to light was that the sample they had been analyzing was a very ancient specimen of human blood, with fragmented and deteriorated chains of DNA, and probably contaminated by both male and female DNAs (the data as they were presented in the article to the benefit of more competent readers are reported in Table 9.2).

Table 9.2 Electropherograms of polymerase chain reaction (PCR) amplifications of the Shroud

Minimum (peak)	Size (Base pairs)	Height of peak	Peak area	Number of scan
77	82.53	21	128	771
85	96.26	41	511	853
90	**107.28**	27	286	903
105	130.33	28	279	1052
105	131.16	35	255	1057
123	161.61	26	351	1231
129	174.14	20	113	1298
130	174.68	20	110	1301
152	212.37	52	418	1527
153	213.04	48	340	1531
155	216.40	98	1258	1551
157	220.63	55	720	1576
159	225.03	59	686	1599
162	230.04	33	319	1625
165	234.56	57	682	1651

Reference: [33].

In 2007, the American scientist F. Tipler [138, pp. 171–187] analyzed again the electropherograms[15] on the DNA published in 1995 and interpreted them not as the product of the contamination of different persons but as the result of the blood belonging to a male with the XX chromosome containing the *SRY* gene.

Tipler supposed that the sequencing of the blood of the Shroud could be the product of a particular parthenogenesis[16] where the *SRY* gene was added to the female X chromosome. In regard to these data, his conclusion was as follows: "They are the expected signature of the DNA of a male born in a Virgin Birth!"[17] Obviously,

---

[15] The electropherogram or electrophoretogram shows the DNA sequence in just one diagram.

[16] Parthenogenesis, from the Greek παρθένος (parthénos), meaning "virgin," and γένεις (ghénesis), meaning "birth," can be considered a form of asexual reproduction that occurs without fertilization. Parthenogenesis can be artificially induced on the egg by treating it with various kinds of physical and chemical stimuli.

[17] With reference to the data of the electropherogram (see [138] on page 186), Tipler says that "the data coming from the Shroud of Turin highlight the 107 (106+1), but there is no trace of the 112, value of the basic couples of the gene. The Shroud of

this is just an interpretation that makes reference to the intervention of the Holy Spirit, even though, however, in the near future it will be important to perform an in-depth analysis of the hypothesis suggested by Tipler.

The team of the University of Padua studied the dusts vacuumed by the Shroud, and not on the samples of blood directly taken from the Shroud itself, finding clear traces of DNA belonging to several persons with genetic diversities. The analysis of the reference haplotype[18] haplogroup[19] effectuated on every discovered sequence had been helpful to make assumptions concerning the ethnic origins of these people. The team identified the presence of human DNA belonging to the haplogroups R0a, U5b, and H1-4:

- The general R0 haplogroup is typical of certain populations of the Arabian Peninsula.
- The branch U5 of the haplogroup U is particularly ancient and common in European populations.
- The haplogroup H is the most frequent one in Europe, but it is present even among some populations of North Africa. For instance, the haplogroup H1 is common in Libya and Morocco and among the Berbers.

In conclusion, in agreement with Professor Barcaccia, the haplotypes found in the traces of DNA have a reference to the haplogroups of ethnic groups belonging to Europe, North Africa, and the Middle East, that is to say, of people living in areas facing the Mediterranean basin. More accurate analyses in the process of development could, in the near future, be more precise and give us specific data concerning those people who in the past—even in a faraway past—have touched, kissed, or come in contact with the Shroud of Turin.

Turin highlights the 105 (106-1), but there is no trace of a 112 basic couple. The X chromosome is present, but not the Y one. This is the evidence of the simplest virgin birth: an XX male generated through a SRY inserted in an X chromosome. This is not something to be expected from a standard male."

[18] The haplotype is the combination of variations in the DNA sequence on a particular chromosome, which tends to be transmitted en bloc.

[19] The haplogroup is a group of haplotypes very different from each other but originated from the same ancestral haplotype.

9.7 Future Developments

Several topics concerning recent discoveries on the Shroud of Turin have been discussed throughout this book. However, the readers must not be led to think that by now there is nothing more to discover around the most important relic of Christianity. Indeed, not everything has been unearthed and today we can say that if we knew the 5% of what could be known about the Shroud of Turin, we would know too much. Everything else is still shrouded in a mystery that science has not yet fathomed.

Many positivist people are convinced, perhaps with a touch of arrogance, that science is able, or will be able in the near future, to explain all what surrounds us, forgetting that science is in itself limited because human knowledge, based on the five senses, is limited and can perceive just part of the complex reality we are part of. After all, the description of physics phenomena is based on theories, that is to say, on a set of equations deduced from basic principles and able to prove experimental results. But theories do not enjoy everlasting fame: for instance, when inconsistencies emerge it may be necessary to introduce (or just to wait for) a new theory. This is the case of the theory of general relativity by Albert Einstein, which was able to explain the anomalies in the orbit of Mercury, when the older Newton's law of universal gravitation was not. In the words of Stephen Hawking, "Any physical theory is always provisional, in the sense that it is only a hypothesis: you can never prove it. No matter how many times the results of experiments agree with some theory, you can never be sure that the next time the result will not contradict the theory. On the other hand, you can disprove a theory by finding even a single observation that disagrees with the predictions of the theory" [77, p. 23].

Moreover, there are not theories for everything. For instance, no theory has been discovered that encompasses, in a set of formulas, the four fundamental interactions (gravitational, electromagnetic, strong nuclear, and weak nuclear). Once discovered, this theory, too, would still be valid until there will be clear evidence to the contrary. Nature with its many phenomena can be extremely complicated, and hypotheses, examples, and theories aim at producing a description of it . . . as much as possible only by using our five senses.

This is the reason why even the Shroud has to be approached with humility, trying to detect all the clues it offers in the attempt to give an answer to open questions ranging from science to theology.

But what are the questions that have not yet been given an answer? What kind of studies will be carried out in the future?

Some 30 years ago, scientists of STURP asked several questions, on a more specific technical level, which to these days have not yet been answered. Some of these questions are shown below.

(A) What is the penetration depth of the relic fabric?

The available data concern the analyses in situ made by R. Rogers[20] in 1978 and the results [60] regarding the double superficiality of the image. However, more specific studies conducted in different areas are extremely important to find an explanation of the formation of the image.

(B) What is the chemical composition of the areas characterized by scourging marks? In what ways do they differ from the areas defined by the presence of blood, haloes, the body image, and the background?

Answers to such a question could explain how scourging marks ended up imprinted in the linen of the Shroud, a point that still today has not found a plausible explanation. Indeed, the visible marks of contusions under the skin layer are not easy to understand.

(C) Do the areas of the shoulders and the calves contain deposits of dust and dirt like the ones of the heels?

Many areas of the Shroud have not yet been studied in such a way to identify the materials (dust or others) they contain. An analysis of this kind will better illustrate the overall framework of the characteristics of the relic and of the man who had been wrapped in it.

(D) How does the Shroud appear in infrared and in visible light?

G. B. Cordiglia [84] had the Shroud and some of its details

---

[20] Private communication to the author of January 24, 2003

photographed in infrared; Accetta and Baumgart [3] made studies using the far infrared.[21] However, this kind of analysis has to be carried out in a more systematic way, trying to find more than one answer to the questions posed by the study of these first photographs.

(E) What are the characteristics of the fluorescence visible in the various areas of the Shroud (e.g., the body, the blood, the stains of water, the texture, the burns. . .)?
The Gilberts [72] conducted some studies, but still, there are points that have to be examined.

(F) On the basis of the different spectra of blood obtained, could their characteristics be explained through age, chemistry, microbiology, the exposition to environmental factors, etc.?
Adler [5] has given some interesting answers that will be analyzed.

In addition to these more technical points, we are left with the following fundamental questions that have not yet found an answer and perhaps never will, considering that they involve not only scientific subjects but also religious ones.

In regard to this, it is important not to forget that the Christian God does not impose Himself; He just makes a proposal, always leaving room for the free will of human beings. Actually, it is up to human beings to decide whether to accept a faith and, thus, what the Shroud tries to convey. In this light it seems easier to understand why the relic of Turin gives us hundreds of clues but never a certain proof that could interfere with a person's faith.

Fundamental questions (which many people would like to answer) concerning the most important relic of Christianity are reported below.

(1) Is the Shroud of Turin authentic?
Before seeking an answer to this question, it is important to better define what is meant with the word "authentic". For some people it simply means that the Shroud is not a European medieval manufacture, as was wrongly believed, but a 2000-year-

---

[21] The infrared extends from the nominal red edge of the visible spectra. The far infrared is a set of wavelengths that are even farther from the red edge.

old manufacture made in the Middle East. At this point other hypotheses could be considered equally valid. For instance, if the Shroud was considered authentic only because of its 2000-year-old fabric produced in the Middle East, even the hypothesis that the double image had been made by an extraterrestrial intelligence could assert its authenticity. Not to mention the idea that it could be the result of a miracle that perhaps occurred in the Middle Ages. It is clear then that the term "authentic" has to be better defined.

Some people believe that the Shroud can be considered authentic only if it had wrapped the body of a man who underwent the same tortures that Jesus did. Others think the Shroud can be deemed authentic only if it had actually wrapped Jesus Christ. Finally, other more demanding people assert it is authentic only if it had wrapped the body of the Saviour, who impressed his image by resurrecting from the dead and releasing a particular energy. Of course, this being the case, the answer goes beyond science, which does not have the means to cope with the phenomenon of the Resurrection, definitely not reproducible.

If with "authenticity" is meant a very old shroud, around 2000 years old, woven in the Middle East, which wrapped the corpse of a man who had been harshly scourged, who was crowned with thorns, and who died on the Cross and could be identified with Jesus of Nazareth, then in this case, an answer can be sought, although not ultimate and with no sure scientific proofs. The first author has personally found his answer after 15 years of studies and intertwining correlations among the numerous clues, not only scientific: the Shroud is authentic.

However, even in agreement with the mind-set of the second author, this answer cannot be certain for those who did not have the chance to study for years this complex matter: it is up to scholars to try to scientifically demonstrate this statement in the near future.

(2) How did the body image form?

Several hypotheses had been considered in order to explain the formation of the body image, which could be compared to the objective data observed on the Shroud. As has been said, the more plausible hypothesis makes reference to a brief but intense emission of electrons: the corona discharge could be responsible for the browning of the linen fibers and the lack of pigment. Are all the issues then resolved? Absolutely not!

This is just a hypothesis that will be verified through future in-depth analyses that may even attempt to reproduce the environment where the phenomenon occurred. For now we do not have definitive answers to the question, for science is not able to reproduce what could just be explained in theory. Indeed, to this day no laboratory in the world has succeeded in reproducing a 1/1-scale copy of the Shroud of Turin, able to show all the optical-physical-chemical characteristics [64] of the original.[22]

Even though it is possible that the body image formed as a result of a very particular event, for now science can provide no answers. Indeed, according to some scientists, to produce such an image a discharge corresponding to a lightning would be needed around the Shroud: nothing of the sort is known to have ever happened, especially inside a sepulcher sealed with a stone. For now we can only try to figure out answers by comparing the many similarities between what has been observed on the Shroud and what is reported in the Gospels, accepting as true what they state: Jesus, who died and was buried wrapped in the Shroud, resurrected. As things stand, the matter is very interesting and open to every scientific solution that can be given. Last but not least, it would be interesting managing to study from a scientific point of view the phenomenon of the *Holy Fire*[23] that every year ignites in the tomb of Jesus in Jerusalem on the eve of Orthodox Easter, perhaps trying to connect it with the body formation on the Shroud. As far as we are aware the Orthodox Church has not allowed scientific studies on the Fire, but as videos on the web show, this fire seems to be very peculiar and does not seem to burn inflammable objects, as a normal

---

[22] The authors are willing to analyze hypothetical possible copies of the Shroud deriving from laboratory tests. If and when the correspondence between the artifact and all the characteristics of the Shroud will be proved, they are ready to change the statement made in the text. As for the past, the first author has already analyzed many artifacts, some of which are very interesting but still very far from the result hoped to be obtained.

[23] *"The great miracle of the Holy Fire occurs every year at noon on Holy Saturday in the Church of the Holy Sepulchre in Jerusalem. Spontaneously the Fire ignites, holding the interest of people who are blessed and honored to be part of the ceremony, making them rejoice and celebrate, and instilling faith in them"* [117].

fire does. Perhaps in the future we will be able to connect the image of the Shroud with this particular phenomenon.

(3) Who is the Man of the Shroud? Is he the Risen One?

On the Shroud there is not the name of the person who was wrapped in it, although some people believed they have spotted letters that could have reference to Jesus of Nazareth. On the other hand, even if both the name and surname of the person were indicated, they could still be a forgery. There is then no certain and unmistakable evidence that could favor the identification of the Man of the Shroud. However, the identity seems to be written in a universal language, a "scientific" language directly connected to the data observable on the relic. Who is able to "read" these elements can find an answer that cannot completely be proved from a scientific point of view, even in the light of what has been previously observed.

In reference to the question concerning the Resurrection, this clearly goes beyond traditional science. Every answer of this sort must then be sought outside the scientific field, perhaps by looking for a correlation between scientific clues and what faith suggests.

Each reader can at this point contemplate the image of the face of the Man of the Shroud and answer the question, "*Who do you say that I am?*" (Luke, 9:20).

Chapter 10

Additional Questions and Answers

The following 24 questions have been selected among those posed to the authors during conferences. For brevity's sake, the answers are not fully exhaustive. The various, at times very complex, aspects involved in each question are not always comprehensively addressed.[1] At any rate, the answers indicate the overall number of issues involved and constitute, in combination with the previous chapters, a further step toward an understanding of the relic, though examining all the topics involved in the Shroud appears to be a difficult task.

10.1 What Can the Presence of Small Coins on the Shroud's Eyes Tell Us?

Greek–Roman mythology contemplates that the souls of the departed place two coins on the eyes of their bodily remains in payment to Charon for the crossing of the Acheron River on their way to Hades.

---

[1] On similar grounds, we do not provide the references for each answer.

The Shroud of Turin: First Century after Christ!
Giulio Fanti and Pierandrea Malfi
Copyright © 2015 Pan Stanford Publishing Pte. Ltd.
ISBN 978-981-4669-12-2 (Hardcover), 978-981-4669-13-9 (eBook)
www.panstanford.com

Studying in 1931 photographs (Fig. 1.9) of the Shroud taken by G. Enrie (Section 2.3) on orthochromatic plates (i.e., plates producing an enhanced black-and-white contrast), F. Filas was the first to detect a prominence on the right eye, possibly related to a coin being placed over it.

Additional studies, conducted by increasing the contrast of the photographs, seem to distinguish a reverse question mark and some letters related to a *dilepton lituus* minted (Fig. 10.1) by Pontius Pilate around A.D. 30.

Figure 10.1 *Dilepton lituus*, coined by Pontius Pilate around A.D. 30, with an imprint of a question mark on a Shroud-like fabric. Courtesy of M. Moroni.

N. Balossino believes to have identified an additional coin, *dilepton simpulum*, coined in the same epoch, on the left eyebrow, but his discovery remains doubtful.

Clearly, if coins were actually placed on the Shroud's eyes, it should be proof that the Man of the Shroud was buried in Palestine around 2000 years ago. But more recent photos of the Shroud show no sign of coins on the Shroud's eyes, which leaves the following question open: Were the signs detected on Enrie's photos an artifact, or is the lack of these signs in the newer photos perhaps due to their lower contrast?

A further aspect should be considered: Jesus was a Jew, not a Roman. Why should those who provided for his burial have followed a pagan tradition rather than the Jewish one? No Jewish soul should cross the Acheron after relinquishing its earthly remains; it just is not a Jewish tradition.

10.2 Are There Any Inscriptions on the Shroud?

Various researchers, including French professor A. Marion, detected the presence of writings around the face of the Shroud Man, namely "Nazarenus," (I)ΗΣΟΥ(Σ) ("Iesoûs") Jesus, or "IC," ΙησούςΧριστός (Iesoûs Christós), Jesus Christ. But the letters emerge from such a complex procedure of contrast enhancing and image processing that not all researchers agree about their actual existence.

Future image processing on photographs directly taken for the purpose will clarify any doubts.

10.3 What Can Be Said about the Historical Journey of the Shroud According to the Study of Pollen?

In 1972 and again in 1978 M. Frei confirmed the supposed historical route (Fig. 2.5) of the Shroud going through Jerusalem, Edessa, Costantinople, Lirey, Chambéry, and Turin. He in fact found (Section 9.2) pollen grains, sampled from the Shroud in sticky tapes, closely corresponding to those typical of the areas in which the relic was exhibited.

After the premature death of Frei, who was unable to complete his interesting analysis, some scholars, such as A. Danin (Professor of Botanics at the Hebrew University of Jerusalem) and U. Baruch, confirmed Frei's conclusions. On the other hand, scholars such as G. Ciccone of the Italian CICAP[2] continued to place Frei's scientific integrity in doubt, also implying fraud in his published results.

Following such accusations, the first author undertook a partial pollen grain analysis using a scanning electron microscope (SEM) for detecting grains in the dusts vacuumed from the Shroud; his results are in agreement with those of Frei, thus disproving suggestions of fraud.

10.4 Is the Blood on the Shroud Human, and Is It of the AB Type?

In 1969 the Pellegrino Commission obtained an uncertain result from the analysis of some red particles taken from the Shroud, because they were not recognized by the tools of the time. In 1978, A. Adler and J. Heller in the U.S. and P. Baima Bollone in Italy independently analyzed the red particles collected from the Shroud (Section 9.4). This time they were identified as human blood. Additionally, Bollone recognized the AB group, but his finding was questioned by the US team, with the observation that blood as ancient as that of the Shroud could furnish false AB results due to its degradation in time.

We may add at this point that tests on blood from relics attributed to Jesus (such as the Shroud of Oviedo and the Eucharistic Miracle of Lanciano) have provided an AB result, too.

Future investigation based on DNA analysis will clarify existing doubts.

---

[2] Comitato Italiano per il Controllo delle Affermazioni sulle Pseudoscienze; in English: Italian Committee for the Investigation of Claims of the Pseudosciences.

10.5 How Was the Man of the Shroud Crucified?

Uncovering the course of action taken for crucifying the Man of the Shroud merely from the information obtainable from frontal and dorsal body images of the Shroud itself is not a simple task.

Crucifixion was an ancient and painfully slow execution technique. It allowed for variations, since at the time of Jesus this form of capital punishment was left to the sadistic "creativity" of the executioners, the Roman soldiers.

The victim in fact could be tied and/or nailed to the cross, and the cross itself came in different shapes and sizes. It could, for example, be formed by a vertical pole (in Latin, *stipes*) permanently standing at the execution place, to which the offender added a beam (called *patibulum* in Latin) horizontally, which he himself had carried on his shoulders. Again, the cross could be produced by binding two wooden beams together. It would be brought by the offender to the execution place, generally outside the city but not too far away from it. In the latter case the cross was hoisted with ropes and set in a hole dug in advance at the execution grounds.

With respect to the Man of the Shroud, no doubt nails were used and discernible wounds are coherent with his wrist and feet being nailed to the Cross. Notably, nails were hammered in correspondence to the wrists, not to the palms of the hands, because the tensile force due to gravity applied to the hand would have torn its muscles and tendons, thus forgoing the nail's function of fastening the body to the Cross. The strength of the nails stuck in the wrists, on the contrary, met with the necessary resistance in the wrist's bones.

Recent studies conducted by the team of M. Bevilacqua also come to the conclusion that the crucifixion technique endured by the Man of the Shroud, as may be observed in the studies of the relic, does not differ from the descriptions encountered in the Bible, and the Gospels in particular. On the basis of the studies mentioned above and in recent specific medical studies and experiments, we can describe the process of crucifixion as follows.

After having carried the Cross (composed of *stipes* and *patibulum*) at the execution place, the Man of the Shroud was probably tied

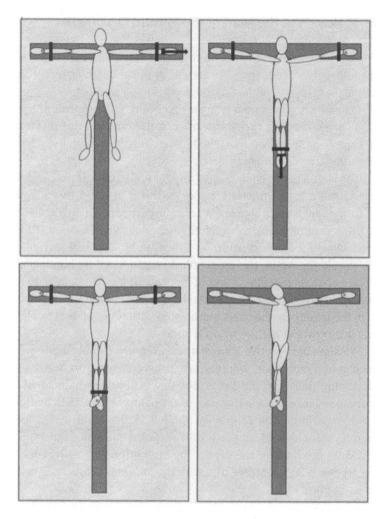

Figure 10.2 Crucifixion procedure. Top left: With the cross lying on the ground, arms are tied with ropes and one wrist is nailed, while the opposite wrist is pulled by ropes. Top right: After nailing the second wrist, the tied feet are pulled by ropes. Bottom left: Having pierced the right foot, the left foot is nailed on top of it. Bottom right: The old ropes are taken away, and the cross is raised by new ones, thus increasing the pull on the wrists, but reducing that on the feet and producing knee bending; it probably represents the final posture of the dead man.

to it while lying on the ground. A nail[3] was stuck in the Destot space of one wrist and then the opposite arm was stretched, the ropes dislocating the bone junctions, so as to position the second wrist in correspondence with the hole previously made on the Cross. If pilot holes were used, the nail heads had to be hammered to fasten the nails firmly to the wood. Hence the second arm was nailed.

The feet were pulled by ropes, dislocating the bone junctions in the process, and the right foot was nailed between the first and second metatarsal bones but the nail was subsequently extracted. A longer nail was then used to pierce the left foot, to be nailed over the right one through the hole previously made in it.

Thus the Man of the Shroud was pinned to the Cross by means of three nails, and the Cross finally hoisted by ropes and fixed in a hole dug in the ground.

10.6 Is the Body Image Formed by Pigment Substances?

The analysis performed by the first author on dusts vacuumed from the Shroud (Fig. 5.8) identifies some pigments on the linen fabric, but these are relatively rare and therefore inadequate to explain any coloration producing the body image as a result.

Parallel analyses on image fibers, again conducted by the first author, definitely confirm on the other hand the absence of pigment or of any other intake substance on the image fibers, in harmony with the results obtained by the Shroud of Turin Research Project (STURP)[4] in their 1978 direct examination of the body image. The image is in fact the product of chemical reactions (oxidation, dehydration, and conjugation) of fibers on the image's surface (Section 1.4).

Among the pigments present in the vacuumed dusts, the first author found particles of lapis lazuli (a blue-colored precious hard stone) mixed with iron oxide particles (which are red), leading them

---

[3]Nails were made of iron with a square cross section. What is commonly known as iron was actually in Roman times a low-carbon steel alloy, or "soft iron." After piercing the wrist, the nail would probably bend on meeting the hard wood of the Cross. A solution was probably found by boring a pilot hole in advance.
[4]Section 2.3.

to suspect external contamination had occurred in the course of the centuries. It is a known fact that in the past centuries artists avowedly made copies of the Shroud, touching the sacred Linen with their paintings, in order to confer to them qualities of the highest order, but physically contaminating the Shroud in the process.

10.7 Why Is the Body Image on the Shroud Not Reproducible?

The Shroud Science Group, numbering over 100 mostly US scholars, has published [64] a list with 187 outstanding characteristics of the body image that nobody has been able up to now to reproduce simultaneously; comprising three-dimensionality, negativity, superficiality, and, in some areas of the image, double superficiality,[5] the lack of pigments, the uniform coloring of the fibers around their circumference,[6] and the presence (Fig. 1.12) of nonimage fibers adjacent to colored ones (Section 1.3 and Section 1.4).

The complete list of remarkable features can be found in (Section A.2) of the appendix or at the following address: http://www.shroud.com/pdfs/doclist.pdf.

The exceptional features of the Shroud image have altogether never been reproduced.

10.8 Was the Height of the Shroud Man in Line with Historical Times?

An anthropometric analysis on a digital superimposition between a copy of the Shroud and a numerical manikin (Fig. 10.3) representing the Man enveloped in the Shroud has determined a height of 175 ± 2 cm (68.90 ± 0.79 in.).

This analysis was performed before the 2002 intervention, in the course of which the relic was elongated by about 8 cm (3.15 in.)

---

[5] Double superficiality indicates that the image is superficially visible on both sides of the cloth, while there is no image at the center of the fabric.

[6] A linen thread is composed of 100–200 fibers with a diameter of about 0.015 mm (0.00059 in.).

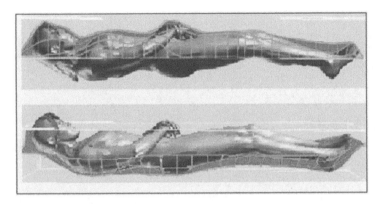

Figure 10.3 Numerical manikin of the Man of the Shroud with the numerical enveloping sheet reproducing the Shroud, used to determine its height.

along one side. Thus, if the same anthropometric analysis were to be taken today, the Man's height would obviously increase to about 178 ± 2 cm (70.08 ± 0.79 in.).

If the Man of the Shroud has stretched out by some centimeters in just a few decades, it is quite plausible that the Shroud would have been elongated in the past centuries, too, perhaps during its public exhibition, and that the original height might therefore have been inferior, measuring approximately 170 cm (66.93 in.).

As Palestinians 2000 years ago were on average 167 cm (65.75 in.) tall, the Shroud Man appears to be perfectly compatible with the average height of the period.

10.9 How Old Is the Man of the Shroud? Is He Really 33 Years Old?

It is quite impossible to determine the exact age of a man from a monochromatic photo in which, for example, the color of his hair is not to be seen. According to a number of forensic analyses, well-pronounced muscles and the absence of a fat belly are not in contrast with the features of a man in his thirties.

Figure 10.4   Oviedo Sudarium.

10.10 What Is the Relationship, if Any, between the Shroud and the Sudarium of Oviedo?

The Sudarium of Oviedo, conserved in this Spanish city, is a linen cloth (Fig. 10.4) of about 53 × 86 cm (20.87 × 33.86 in.). It has many bloodstains consistent with the head of a tortured man being wrapped up in it. In accordance with tradition, and with a number of scientists, it is believed that the Sudarium has been used for wrapping up the head of Jesus Christ as He was taken down from the Cross.

The marks on the Sudarium are consistent with those of the Man of the Shroud. Many other marks put the two cloths in relation, including the holes produced by the crown of thorns that appear to be visible in the Sudarium.

The Spanish relic, however, does not disclose any body image. It cannot therefore straightforwardly be stated that the two cloths were used for the same person, even if the bloodstains have been analyzed to be of human origin and, according to Bollone, precisely of the AB type. Future DNA analyses will confirm whether or not a correlation between the two linen sheets is possible.

Figure 10.5   The Manoppello Veil.

10.11   What Is the Relationship with the Holy Face of the Manoppello Veil?

The Manoppello Veil, measuring 17×24 cm (6.69×9.45 in.) is a handkerchief made of fine linen or byssus[7] that shows an image (Fig. 10.5) attributed to Jesus Christ. It is conserved at Manoppello, a small town in Italy near Pesaro (in the Marche Region).

---

[7]See footnote 3 on p. 167.

The face image painted on the Veil has some particularities that cannot easily be attributed to a human painter.

Importantly, at least two different superimposed images can be detected if the illumination changes the incidence angle. No painting technique is known to produce this result.

As the Veil is sealed between two glass panes it is not possible at the moment to extract physical samples for scientific analysis. Even so, P. Baraldi of Modena (Italy) University, taking Raman tests (Section 6.3) through the glass, has claimed that the pigments composing the image are probably not from any substance known on Earth. From the scientific point of view it is clear that the issue requires further investigation with more specific analyses once the object is taken out of the reliquary.

Some details, such as bloodstains, can be correlated to those of the Shroud. Others, on the contrary, lead to suppose that the image was impressed at different moments in the Veil and in the Shroud. In fact, while the Man of the Shroud appears to have his eyes closed, the face in the Veil has its eyes open. The swelling of the cheeks appears instead to be a common feature in the two images.

10.12 What Is the Relationship with the Tilma of Guadalupe?

The famous Tilma of Guadalupe, also called "Our Lady of Guadalupe," is conserved in Mexico City. It is a cloak belonging to St. Juan Diego, on which in 1531 the image of the Mother of God was miraculously impressed. The Bishop of Mexico City, who was present on the spot, witnessed the miracle.

The image (Fig. 10.6) appears to have been painted by a non-human artist but underwent many different restorations through the centuries, so carrying out a scientific analysis on the original painting is a complex task. The fabric, as in the case of the Veil, is enclosed in a reliquary, kept unopened for some decades, preventing direct access to the object. The back side of the Tilma contains an image, which, too, is quite deteriorated but of great scientific interest because it was never subjected to restoration.

Figure 10.6 Our Lady of Guadalupe.

Finding some correlation between the images on the Tilma and on the Shroud is not so easy at a first sight: the image on the Tilma is formed by pigments, while the image on the Shroud is not. The figures represented in the two cloths are different—in one fabric the Mother; the Son in the other. Both images seem to be not made by human hand.

A more in-depth analysis may instead reveal a possible closer correlation between the images. The first author has a rare photograph of the back side of the Tilma kindly furnished by the Mexican Professor A. Orozco and, studying it in detail, has discovered two red spots indicating bloodstains in correspondence to the forehead and the wrists. These red spots are not visible on the frontal side because the image was subjected to restoration. They closely correspond in position to the clearly visible reverse "3" on the forehead and to the wrist injury on the Shroud image, but appear to have nothing in common with the image of the Mother of God.

Setting aside science, the correspondence could be interpreted as the will to show the strict correlation between Mother and Son through the two different images, a sign that while the Son is the Redeemer, the Mother is the Redemptrix.

10.13 What Differentiates the Shroud from Other Images Not Made by Human Hand?

In addition to the Byzantine icons that sometimes reproduce the so-called Christ Mandylion or Acheiropoietos (not made by human hand), a few other images may be considered as not made by human hand. Among them we may mention the Shroud of Turin, the Tilma of Guadalupe, the Veil of Manoppello, and some handkerchiefs of St. Pio from Pietralcina. One of the latter reproduces on one side Father Pio's face and a Christ-like face on the opposite side (Fig. 10.7).

All these images, however, differ from the Shroud because the double body image impressed on it is devoid of pigments of any kind. The Shroud's color depends only on a chemical reaction of the outermost layer of the linen fibers compositing it.

Traces of human blood are also imprinted on the Shroud but are not present in the other images, with the exception of the Sudarium of Oviedo, which has no images.

From the experience of the first author, who studied the Acheiropoietos images in depth, it seems that God does not use a particular technique for producing these images, but makes the best use of the natural sources currently at His disposition in that instant.

Figure 10.7 The St. Pio handkerchief: on the left is a Christ-like face image and on the right is a St. Pio face image visible on the opposite side.

10.14 What about the 1988 Radiocarbon Dating? Three Accredited Laboratories Published Their Findings in the Famous Journal *Nature*, and 24 Scientists Signed the Document. How Can They All Be Wrong?

There are some facts to consider. M. Tite, the group leader, has not answered recent queries and clarification requests raised by some researchers, the first author included. The protocol proposed for an accurate sampling (Section 4.2) was not followed in 1988; samples were taken (Fig. 4.2) in a darker, and thus more contaminated, area of the Shroud.

Apart from these notes, from a scientific point of view it must be observed that the radiocarbon dating method (Section 4.1) does not always furnish reliable data. Expert scientist W. Meacham wrote that a nonnegligible percentage of results are wrong principally due to external contamination of the sample. For example, as mentioned in Section 4.3, analysis on the Egyptian mummy n.1770 preserved in the Manchester Museum furnished different dates for bones and bandages—the latter resulting 800–1000 years younger than the former because of external contamination. Another example refers to the dating of young shells (*Melanoides tuberculatus*) that were classified as about 27,000 years old due to external contamination.

In the case of the Shroud, a recent robust statistical analysis by M. Riani et al., also published [120] in an important scientific journal, demonstrated the inconsistency of the 1988 result, trying to explain this fact with an external contamination that could have altered the percentage of carbon isotopes. In particular, a nonnegligible trend in the 1988 result was detected (Fig. 4.5), which as a consequence can increase the uncertainty of the result, even by thousands of years. This important statistical analysis also revealed that the laboratory in Tucson in Arizona only tested one of the two Shroud samples furnished for dating purposes. This fact was subsequently confirmed by the laboratory itself (Section 4.3).

Therefore, even if the three laboratories have made a reliable analysis in 1988, problems connected with external contamination may have played an important role, thus producing an unreliable result. To prove as much, we submit at length in this book the dating analyses by the three independent methods, demonstrating the incongruence of the 1988 radiocarbon dating result.

10.15 Does the Energy That Presumably Produced the Body Image Come from the Enveloped Body?

The body image is unaccountable from a scientific point of view, the most probable explanation being attached to an energy burst (which could also be electrical) coming from the human body enveloped in the Shroud.

Some scientists associate such a burst of energy with Resurrection, thinking to a sort of "photograph of the Resurrection." But clearly at this point, as the Resurrection is a scientifically irreproducible phenomenon, any kind of scientific evaluation must fall way short of this statement.

If we can quite easily infer the origin of the energy at the instant in which the body image was formed, we are not in an equal position in understanding the origin of the energy at the instant immediately preceding. In the hypothesis of an energy burst of the electric type (the so-called corona discharge effect), theoretical models and experimental results (Section 1.5) both show that the surface of the human body enveloped in the Shroud must have been electrically charged.

It can therefore quite easily be assumed that the energy in question came from the human body at the moment the image formed itself, but this theoretical model is not able to foresee if this large amount of energy was produced by the body itself or if it derived from an external energy source that previously interacted with the whole system. Similarly, science can explain quite well what happened during the Big Bang but is unable to describe what happened an instant before.

One of the hypotheses proposed to explain the large amount of energy involved refers to the electric energy produced in nature by lightning entering the sepulcher. Setting aside the objective difficulties of explaining how lightning could penetrate a sepulcher sealed off by a large stone, the electric energy produced by the enveloped body could be justified as the consequence of an external event, lightning, that entered the sepulcher and somehow charged the human body electrically.

10.16 Is the Conservation State of the Turin Shroud Compatible with Its Presumed Age of 2000 Years?

The Shroud presents at close reading some holes due to the 1532 fire (Fig. 1.3), some water stains, and other signs of aging but on the whole appears to be a well-preserved fabric. Can a linen tissue reach the age of about 2000 years? The answer is positive. Two facts must be remembered.

First of all, linen is a fabric highly resistant to aging; proof of the fact are the many linen bandages used to wrap Egyptian mummies, relatively well preserved even after 6000 years, and the fabrics of the Kha tomb in the [Egyptian] Museum of Turin, showing relatively few signs of damage after 3500 years. If a linen fabric is maintained in a relatively dry environment it can be preserved for several millennia.

Secondly, the Shroud has always been considered a very important, if not the most important, relic of Christianity and therefore properly conserved. It has been exposed to various conditions but always with great care. In addition, it has been long sealed in suitable reliquaries, which can easily account for its relatively high degree of conservation.

10.17 Was the Burial Procedure Resulting from the Shroud in Line with the Jewish Tradition of Its Presumable Age?

In the Gospel of St. John (19,40) it is reported that the body of Jesus was *bound in linen cloths with the spices, as the custom of the Jews is to bury.* According to experts in Jewish history, all what results from studies of the Shroud is in agreement with the Jewish tradition, but some facts point to a burial procedure reserved for priests of high rank or for kings. For instance, a large quantity of precious spices in powder (100 lb.) was used to at least partly cover the stone on which the Man's lifeless body was lowered, a procedure reserved for very important figures.

First, half the Shroud was laid down on a bed of spices and the corpse lain over it (Fig. 1.22). Then, more rolls of linen impregnated with spices were probably placed around the corpse (Section 1.8). Lastly, the body was covered with the other half of the Shroud.

Since the ritual took place on a Holy Friday evening, it was decided to complete the burial the day after Holy Saturday. This is why the burial procedure may appear only half-finished.

The Gospels tell that the women went to the sepulcher to complete the burial on the morning of Sunday but did not find Jesus' body. The Shroud indirectly confirms this account, because nobody could have remained wrapped up in the linens for more than 40 hours without signs of putrefaction being detected.

In short, nothing in the Shroud seems to go against the Jewish tradition of its age.

10.18 Was the Use of a Shroud Common at the Time?

The answer is yes, it was. At the time Christ was living in Palestine Jewish burial practice consisted of wrapping the corpse in a shroud

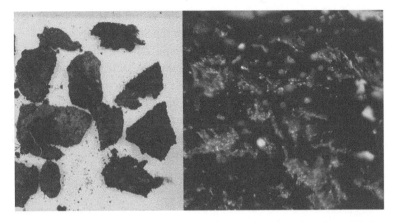

Figure 10.8 Fragments of the Akeldama shroud on the left and zoom of Z-twisted threads on the right.

made of wool or linen; but not all bodies were buried this way. For instance, the bodies of crucified men were normally left to raptors, but in the case of this relic, if the body of the Man of the Shroud was in it, the Gospels relate that Joseph of Arimathea requested, and obtained from Pontius Pilate, the corpse of Jesus.

Different shrouds dated around 2000 years ago have been discovered in the Palestinian area. Some, like the Akeldama shroud (Fig. 10.8) recently found near Jerusalem, are in very poor condition due to the relatively high humidity of the area. Other shrouds, instead, discovered in drier areas such as Kumran or Masada, are very well preserved.

Two facts are noteworthy. Firstly, shrouds are relatively rare to come across, because generally they decompose with the corpse. Secondly, no shroud except the Shroud itself shows an image of the body that was wrapped in it.

10.19 Could Science Explain Everything of the Shroud Tomorrow?

This is obviously an impossible prediction to make. Science has made great strides in general and should progress even more in the near future. It may therefore safely be assumed that a lot of new discoveries will be made in relation to the Shroud, too.

But will human beings be able one day to know everything about the Shroud? And will human beings be able one day to know everything about the Universe? At this point, we must be reminded that science is limited because it is guided by human beings who themselves are limited. On such account, it is quite hard to imagine how human beings will grasp all its secrets, considering the Shroud stems from God, at least in the opinion of the first author.

10.20 Why Do the Gospels Make No Mention of the Body Image Impressed on the Shroud?

The formation of an image on a funeral shroud truly is a most uncommon fact, hence likely to obtain wide publicity. This might be an obvious observation at first sight, especially for twenty-first-century people, whose culture is based to a great extent on images. Not so in the first century, when such a culture was not as widespread, and a not very distinct image appearing on a cloth could have easily been taken for a stain.

Besides, a shroud enveloping a corpse, considered impure by Jews, was itself considered impure and, as such, to be hidden.

Last but not least, it must be remembered that if, as hypothesized by many scientists, the image was produced by an electric field through the so-called corona discharge[8] effect, the latter effect produces a latent image that develops with time. Experimental results show that it is necessary to iron the linen at temperatures of about 200°C to "develop" the latent image in order to see it immediately after exposure to the electric fields.

Following the corona discharge hypothesis, the latent image was not visible at the beginning. It developed in the following years and for this reason went unmentioned by the Apostles in the Gospels.

---

[8]See footnote 23 on p. 28.

10.21 How Come That the Body Image Is Still So Well Defined after All These Centuries?

We have seen in the answer to question 16 (Section 10.16) that if properly preserved (in relatively dry environments) linen is time resistant because it is mainly composed of cellulose, a material that shows few signs of deterioration in time; studies related to mechanical dating have proven this as a fact.

The body image is created by a premature aging of the linen fibers (the chemical reactions involved are oxidation, dehydration, and conjugation) that are relatively time resistant, too (Section 1.4). If the body image was instead made of pigments, these could very easily become detached in time, thus leading to quick deterioration of the body image.

Also noteworthy is the contrast of the body image with the background, which tends to become yellower in time, thus reducing the image's readability.

We have seen in the answer to question 20 (Section 10.20) that the body image "developed" in time. It reached maximum contrast conceivably during the Byzantine period, because after that the background yellowing began to reduce the contrast. This might explain why Byzantine coins (Chapter 3) show details of Christ's face that are scarcely perceptible on the Shroud's image today.

From the Byzantine period up to the present day, while the color of the body image remained practically unaltered, the background turned yellower, thus reducing relative contrast. In future centuries the color in the body image will remain roughly stable, but the background will turn more and more yellow to the point of practically wiping out the body image.

10.22 Does the Image on the Shroud Agree with the Hypothesis That It Is the Image of Jesus?

Consistency between the Shroud and the Passion of Jesus Christ, both prominent features in the Bible, especially the Gospels, and also in tradition, is so high and touches so many points that to number them all would be out of the question. A hypothetical artist aiming to reproduce all the congruencies between the Shroud and the Passion would encounter a great deal of difficulty. Section A.2 of the appendix reports some of these congruencies in a list of evidence.

For example, the Shroud shows a man crucified with nails, as reported by Christianity, but notably, that man was crowned by thorns. Nailing was not uncommon in Roman times, but crowning with thorns was a unique or extremely rare procedure reserved for alleged kings. The fact that the Shroud Man was crowned with thorns supports therefore the idea that the Man of the Shroud is Jesus Himself.

Various probabilistic calculations have been conducted (on the question of congruity). The first author carried out one such calculation with E. Marinelli. The result seems very clear: with 100% probability and negligible uncertainty the Man of the Shroud is Jesus Christ Himself. But can probabilistic calculations definitively solve the question? No, they cannot. Whether the Shroud enveloped Jesus' body or not will probably remain an open question forever because many clues do not build up a proof. Furthermore, ascertaining that the age of the linen sheet is compatible with the period in which Jesus lived in Palestine does not answer the question about who has been enveloped in it.

Outside science, there is another aspect to consider. There are a very great number of clues in favor of the hypothesis that the Man of the Shroud is Jesus, and none against it, but there are no sure proofs. Usually, when studying something in sufficient depth, some kind of evidence comes out. Why in our case does no (indisputable) evidence emerge? The first author considers this fact the best demonstration that the Shroud image originates from God. God proposes but does not impose Himself; thus, everyone can decide according to his or her free will. There are many clues in favor of the hypothesis that the Man of the Shroud is Jesus, but everyone is ultimately required to take the last step alone.

10.23 How Can Be Explained to the Pilgrims the Debate on the Shroud Offered by Media?

In the opinion of the first author, we live in a period of rationalism and anti-Catholic positivism. The media are very powerful and often spread scientifically unverified news about the Shroud. If on the one side people might express a general disinterest in the relic, on the other side the fact that millions of pilgrims are booked to see this relic indicates in fact that the Shroud acts on people in different ways: it might simply be speaking to the heart.[9]

10.24 Finally, What Can Be Said about the Shroud?

From a scientific point of view, the Shroud is a linen sheet that enveloped a tortured man who died on a cross. The very particular double body image impressed on it is neither explainable nor reproducible, notwithstanding the numerous attempts made to reproduce it in different ways.

The point of view of the first author is the following. Many scholars have recognized Jesus Christ in the Man of the Shroud,and some suppose that the body image formed during a burst of energy related to the Resurrection. Others, on the contrary, perhaps irritated by the Easter message implied by the relic, prefer to insist that the Shroud image is man-made, also setting aside scientific evidence contradicting the latter opinion.

In fact we may assume that some scholars use a little science to walk out on God. Quoting Pasteur: "A little science estranges men from God, but much science leads them back to Him."

---

[9]The Shroud speaks directly to our hearts, if we are so humble to listen to its soft gentle "voice." The great number of people who attended the previous Shroud exhibition show that many humble people are able to hear its "voice." As an illustration, the first author read a cardinal, custodian of the relic, who wrote that "we do not know who was wrapped into the Shroud," but who also, talking with a humble custodian of a church about the exhibition of the Shroud, was impressed by his words, "As far as I read the newspapers and watch the TV, for sure the Shroud is authentic: as soon as I look at it, my heart tells me it is true."

The Blessed Sebastian Valfrè, who restored the Shroud in 1694, wrote, "The Cross received Jesus alive and returned Him dead; the Shroud received Jesus dead and returned Him alive."

The first author, leaving a scientific perspective and after long years dedicated to studying the Shroud, reaches the following conclusions on the subject. The Shroud is a particular "photograph" of Jesus Christ in His Resurrection, displaying the signs of the tortures He freely suffered for us all to be redeemed. It is the only "photograph" of Himself He allowed us to admire. It is addressed to the many doubtful persons and to those whose Christian faith is weak, reminding them that Jesus Christ has lived, and is living now, among us, and will be waiting for us at the end of our earthly life. In fact the Shroud has been especially given for persons with the hope that many others, pushed by the scientific interest aroused by this "impossible image," come closer to Him and love Him better.

The point of view of the second author is the following. There are people, to whom the second author, too, belonged in the past, who believe that the ^{14}C dating produced in 1988 (Chapter 4) has solved all the problems around the Shroud, turning it into an object devoid of any scientific interest. Furthermore, they believe that, thanks to science, the 1988 result undermines its religious value.

But the problems are not solved at all, even if a different dating result has been obtained. In fact the dating of the Shroud is only one piece of a variegated mosaic in which joint research with the first author has led to novel contributions. For instance, the body image formation process is still scientifically unsolved. This constitutes in itself more than sufficient evidence, capable of stimulating the objective and unprejudiced curiosity of the scientist. Compared to his previous opinion, a deeper knowledge of the relic has induced a change of mind in the second author, who nevertheless still remains altogether indifferent to any religion issue connected with the Shroud.

It must be noted that the dating procedure described in Chapter 7 refers to an undated Roman–Judaic sample. Only at the end of the research, developed within a degree thesis, the exact origin of that generically defined Roman–Judaic fabric has been revealed to the second author, opening him to a fascinating world better to be studied in depth and not only from the mechanical point of view.

Indeed, to seriously address such a complex object as the Shroud, all its basic characteristics have to be known.

As sustained in the previous chapters, the Shroud is definitely a much more complicated issue than might appear at first sight—an indisputable fact from a scientific point of view, worthy of being better known, this being the ultimate aim of our book.

Conclusion

In 1898, with the first photograph taken of the Shroud, a remarkable scientific development aimed to unveil the still concealed mysteries of the Turin relic began. Since then, interesting studies have been developed, which culminated in 1978, with the in-depth research carried out by the Shroud of Turin Research Project (STURP), which led in the years after to many confirmations, even partial, of the authenticity of the Holy Linen.

In 1988, the radiocarbon 14 testing, which dated the Shroud back to the Middle Ages, dampened scholars' enthusiasm and made them wrongly believe that the relic was a fake, just a medieval artifact produced by an artist who skillfully reproduced something mysterious.

Since that date, several scientists have tried to understand more about the radiocarbon result, the only one clashing with the plenty of evidence in favor of the sheet's authenticity, making various suppositions in order to explain the mistakes made during the radiocarbon 14 test in 1988. In 2005 the American chemist Ray Rogers was the first to publish in a scientific review an analysis that demonstrated that the Shroud was much more ancient than the assigned date given by the radiocarbon test, but he could not determine the age with enough accuracy.

The research project financed by the University of Padua allowed developing of different methods of alternative dating through which the Shroud has been dated again. In fact, the two chemical methods based on spectrometry, Raman and Fourier transform infrared (FT-IR), and the multiparametric mechanical test, combined together by an arithmetic mean, date the Shroud back to 33 B.C., with ± 250 years' uncertainty and within a 95% confidence level. It is important to observe that the only period matching with all the three methods is just the first century A.C., the time in which Jesus Christ lived.

At this point, the question asked in Chapter 9, Is the Shroud authentic? seems to have found an answer, but it is not really just like this. It has been shown that the term "authentic" has many interpretations and it is not easy to find a clear answer.

According to the authors, the two chemical dating tests associated to the mechanical test confirm that the Shroud is coeval to the age in which Christ lived. In addition, the numismatic evidence corroborates the Shroud's existence before the period between 1260 and 1390 A.C. established by the wrongly performed ^{14}C testing, because it shows that in 692 A.C. the Shroud was taken as a model for the coins depicting the face of Christ.

However, from a scientific point of view, it would be hard to reach rushed conclusions as far as the relic's authenticity is concerned, in the sense that it did envelop or did not the body of Jesus Christ. In fact, also the hypothesis, even if extremely improbable, that a more recent artist would have reproduced the image of the Redeemer not on a cloth contemporary to his age but on a more than 4 m (13 ft.) long sheet dated back to the first century A.C. and from then still well conserved (supposing that he could have found one) should be rejected with scientific certainty.

Therefore, scientifically speaking, there is still a huge amount of work to be done, and the Holy Linen gives a very strong push to carry on the investigations. If today scientists are aware of 5% of the things that could be known about the Shroud, the remaining 95% has still to be discovered.

Beyond this, there is another aspect to be considered; a believer should not need to recognize Jesus Christ in a sheet: "Blessed are those who have not seen and yet have believed!" (John, 20,29). This is a meaningful sentence Jesus told to Thomas, the apostle! The real Christian should believe just by faith.

This point has been raised also from many clergymen convinced of the almost uselessness of the Shroud for the real Christian, since they affirm that it is not strictly necessary for believing in God. Even better, the true believer believes by faith and not on the basis of the Shroud. If then, for these reasons, issues about the relic's authenticity should be irrelevant for a believer, they should be even more for people who do not believe.

This is true, but if the Shroud were authentic, then why has it been given to us? And why has it been conserved for 2000 years after various vicissitudes, fires, and accidents? The answer is simple: it has been donated to us because also people with a weak faith, a lot nowadays, and who want to put their finger into the wound[1] like Thomas before trusting in the Lord and especially in the Resurrection could believe. After all, also John the apostle, well educated by the Master and put in front of the facts of Mount Tabor, believed in Resurrection only after having seen the empty Shroud in a very particular position into the sepulchre ("... he saw and believed," John, 20,8).

In Chapter 1 we stated that "lots of hints support tradition ... science does not have the 'case solution' yet and yet the Man depicted on the linen sheet does not have a name" and it has to be remembered also now that new evidence has come up.

This is the scientific conclusion we cannot go beyond, and maybe it will never be possible to, because it will not be easy to give a name with full scientific certainty to a figure of a man depicted on a sheet. In fact, even if, for example, in the future it would be possible to read with certainty the name "Jesus Christ" on the Shroud the doubt about who wrote this name and with what intention would surely immediately rise.

The scientific dissertation therefore stops here, in the awareness that science has limits because human knowledge has as well.

Whereas the second author, who has a recent experience in Shroud studies, thinks it is enough to stop here, the first author, aware of going beyond the scientific field, tries to continue a little bit further, taking into account more general aspects based on considerations reached in more than 15 years of in-depth studies on the most important relic of Christianity.

This "evidence" related to dating, together with the many others described in this book, is the basis on which today we can demonstrate (not only on a scientific basis) that the Shroud is

---

[1] "Unless I see in His hands the imprint of the nails, and put my finger into the place of the nails, and put my hand into His side, I will not believe" (John, 20,25).

authentic in the sense that it enveloped the body of Jesus, dead and resurrected in Jerusalem in 30–33 A.D. after being crucified on the Calvary, and recognize the image of our Redeemer imprinted on the Holy Linen.

About this "evidence," someone could affirm that it is:

- "Less," because, for example, the dusts of the Jerusalem ground just identified on the Shroud could refer to another Middle Eastern city exposed to desert winds. As it has been shown, a certain proof that could determine a definitive judgment does not exist.
- "Many," if we think of more than a hundred clues on the Shroud that confirm what is written in the Bible and what is told by tradition.
- "Sufficient" to give that "sign" that many people search for having a basis for their Christian faith.

The first author's belief, reached after accurate scientific research at the microscopic level on Shroud samples, together with theological research, is that the body image imprinted on the sheet could be attributable to a short but intense burst of energy, probably connected to an electric field (corona discharge) released by the enveloped body: basically the only "photograph" that Jesus Christ left to humankind. A "photograph" that reminds not only of His pains voluntarily suffered for men's redemption during His Passion but a "photograph" that also was taken exactly in the moment of His passage to eternal life—hence explaining that hieratic and calm face not typical of a man who suffered all the tortures highlighted by His scourges.

At this point, it does not seem enough that the reader, interested in this issue, comes to recognize the Man of the Shroud. This will be for many only the starting point to get closer to Him and to try to understand why 2000 years ago was left only that one "photograph."

Why this all has been donated to humankind? Is this maybe a sign that has to be properly interpreted? Is it perhaps necessary to read the huge pains suffered by this Man as the condition needed for human redemption? Is this maybe the proof of what is going to happen after death, the resurrection of the flesh? In summary, does the Shroud indicate that every man perhaps should change his life objectives, thinking that he is just a pilgrim on earth to become a citizen of Heaven?

Appendix

Notes for More Interested Readers

In this appendix the most interested readers can deepen some arguments. It is not meant to be read page after page as a normal chapter. In fact it has been organized as a list of independent issues to integrate some aspects covered in the main text. Thus each section has to be read as a supplement to the chapter to which it refers.

A.1 Cap. 1: Exhibition of the Holy Shroud in 2015

On December 4, 2013, the Archbishop of Turin, Monsignor Nosiglia, has given the following press release: *I have the joy to announce that the extraordinary exhibition of the Holy Shroud will be held in the Cathedral of Turin, in 2015.*

So the most important relic of Christianity will be exposed between April and August 2015 in conjunction with the celebrations for the bicentenary of the birth of St. John Bosco (August 16, 1815, Castelnuovo).

Nosiglia also said, *We are confident that on this occasion Pope Francis will come to pray before the Sacred Linen and to honor St. John Bosco, thus sealing an extraordinary year for our ecclesial and civil communities.*

How many were the exhibitions of the Turin Shroud (TS) in the past? The answer is not simple, because there are not clear historical records of all the relic's exhibitions; then we have to specify what is meant by the term "exhibition," because different meanings can be assigned to it, which are clarified below by dividing them into three categories:

(1) Public exhibitions, mostly referred to as "solemn," during which the relic was shown to the people for several days.
(2) Recognitions carried out by groups of people, including groups of scientists who had the task of analyzing some physicochemical aspects of the sacred Linen. For example, the first author of this book, along with other scholars, had the good fortune to attend a reconnaissance in 2002 after the intervention that carried off the patches of the sixteenth century (Chapter 1).
(3) Recognitions carried out by a smaller number of people. For example, to check the state of preservation after the fire of 1997, a few experts in the presence of the Guardian, Cardinal Saldarini, had to recognize the integrity of the relic.

In agreement with what it is reported on the www.ecodito-rino.org and www.piemontesacro.it websites, here is the history of the exhibitions that occurred in the town of Turin. As we mentioned in Chapter 2, before it was brought to this city in 1578, the Shroud made a long trip, more or less well documented, from Jerusalem to Edessa, Constantinople, Athens, Lirey, and Chambéry (Fig. 2.5), and was transferred to various other places to shelter the relic from possible looting during hostilities between cities and countries; for example, during World War II it was housed in the sanctuary of Montevergine (Avellino).

Only the exhibitions of the first type in Turin are considered below, with the indication of the reason of exposure, when known.

The XVI Century

- 1578. St. Charles Borromeo, Archbishop of Milan, venerating the Holy Shroud in Turin to fulfill a vow
- 1582. Third pilgrimage of St. Charles Borromeo in Turin
- 1585. Wedding of Carlo Emanuele I and Catherine of Austria
- 1586. Birth of Philip Emmanuel
- 1587. Birth of Victor Amadeus

The XVII Century

- 1604. Exposition
- 1613. Exposition
- 1614. Alleged exhibition
- 1620. Marriage of the Duke Victor Amadeus with Christine of France, which took place the year before
- 1623. St. Francis de Sales opening the reliquary and presenting the Shroud to the people
- 1625. Wedding of Vittorio Amedeo II
- 1639. Pilgrimage of St. Jane de Chantal
- 1642. Exposition to celebrate the peace of the Savoy family, putting an end to internal wars
- 1661. Exposition
- 1663. Marriage of Duke Carlo Emanuele II with Francesca d'Orleans
- 1664. Pilgrimage of Father Dominic from St. Thomas
- 1665. Marriage of Duke Carlo Emanuele II with Maria Giovanna Battista of Nemours
- 1668. Exposition
- 1672. Exposition
- 1674. Exposition commissioned by Duke Carlo Emanuele II, *to solemnize with the usual Mercy and Devotion the festivity of the most sacred Shroud*
- 1683. Exposition
- 1685. Marriage celebrated the year before of Duke Vittorio Amedeo II with Anne of Orléans
- 1694. Transportation of the Holy Shroud in the Guarini Chapel

The XVIII Century

- 1706. Siege of Turin and transport to Genoa
- 1717. Exposition
- 1720. Union of Sardinia to the Savoy States
- 1722. Wedding of Prince Charles Emmanuel III with Louise Anna Cristina Sultsbach
- 1724. Probable exhibition for the wedding of Charles Emmanuel III with Polyxena Christina

- 1730. Birth of Princess Maria Felicita
- 1735. Exposition
- 1737. Carlo Emanuele III deciding the exhibition
- 1750. Wedding of Vittorio Amedeo III with Clotilde of France
- 1769. Arriving in Turin of Emperor Joseph II
- 1775. Marriage of the Prince of Piedmont, Carlo Emanuele IV, with Anna Maria Clotilde
- 1785. Exposition
- 1798. Carlo Emanuele IV decreeing the exhibition in his apartment
- 1799. Monsignor Buronzo of Signa, Custodian of the Shroud, decreeing its exhibition

The XIX Century

- 1804. Exposition for Pope Pius VII on the road to the coronation of Napoleon I
- 1814. Return of Vittorio Emanuele I
- 1815. Return of Pius VII from the captivity at Fontainebleau
- 1822. Ascent to the throne of Charles Felix
- 1842. Wedding of Vittorio Emanuele I with Maria Adelaide
- 1868. Wedding of Umberto I with Princess Margaret
- 1898. Several centenarians and exhibitions of sacred art

The XX Century

- 1931. Wedding of Umberto II and Maria José of Belgium
- 1933. The Holy Year
- 1973. First television exhibition
- 1978. Fourth centenary of the transport of the Shroud from Chambéry to Turin
- 1998. Five hundredth anniversary of the consecration of the cathedral of Turin and centenary of the relic's photography in 1898

The XXI Century

- 2000. The Great Jubilee.
- 2010. Exposition to find "hope and confidence" in the image of the Passion, Death, and Resurrection of Christ (meaning assigned by the Guardian, Cardinal Severino Poletto). The

exhibition was also performed to show the Shroud after the intervention of 2002, which, among other things, eliminated the sixteenth-century patches on the holes produced by the Chambéry fire.

- 2013. Television exhibition desired by Pope Benedict XVI.
- 2015. Bicentenary of the birth of St. John Bosco.

As we can see, the Shroud was exposed 5 times in the last quarter of the sixteenth century, 18 times in the seventeenth century, 15 times in the eighteenth century, 7 times in the nineteenth century, five times in the twentieth century, and already 4 times from 2000 to 2015, perhaps to demonstrate the increased interest of the public for the Turin relic or the intention of the Church to call back to sacred things people distracted by other more material interests.

A.2 Cap. 1: The 187 Peculiar Characteristics of the Shroud Image

We report the whole text of an important paper [64] written in 2005 by 24 scientists of the Shroud Science Group in which 187 peculiar characteristics of the Shroud image are described. For space reasons, we excluded the references section, but the reader can easily download this work from http://www.shroud.com/pdfs/doclist.pdf.

Evidence for Testing Hypotheses about the Body Image Formation of the Turin Shroud

Giulio Fanti, Barrie Schwortz, August Accetta, José A. Botella, Berns J. Buenaobra, Manuel Carreira, Frank Cheng, Fabio Crosilla, R. Dinegar, Helmut Felzmann, Bob Haroldsen, Piero Iacazio, Francesco Lattarulo, Giovanni Novelli, Joe Marino, Alessandro Malantrucco, Paul Maloney, Daniel Porter, Bruno Pozzetto, Ray Schneider, Niels Svensson, Traudl Wally, Alan D. Whanger, and Frederick Zugibe

Foreword

This paper has been written in honor of the lamented Raymond Rogers who first proposed this work and dedicated many hours of his life to improve this collection of information; he wrote:

> *No matter what the truth is about the Shroud, it is a fascinating*
> *study. It can be studied according to the rigorous Scientific Method,*
> *and it is too bad that so many wild flights of fancy have destroyed*
> *the credibility of the studies. Maybe we can restore some science to*
> *the discussions.*

The Shroud Science Group and the present paper are a first effort to realize Rogers's hope.

Summary

This paper is the first document, still in progress, that derives from a very wide discussion on the Yahoo! Shroud Science Group and has the aim to present all the evidence detected on the TS that can be useful for a further discussion about the problem of the body image formation.

Many hypotheses about the image formation have been proposed, but up to now, none, scientifically testable, satisfies simultaneously all the facts detected on the Shroud.

The group has the aim to consider in depth all the possible hypotheses proposed or to improve some others in order to determine if some mechanism, more or less complicated, is able to explain all the many peculiarities of the Shroud.

This aim is not simple because the group has no access, for the moment, to the results of the new tests on the Shroud made under the guidance of Prof. P. Savarino, scientific consultant of the Custodian Card. S. Poletto, in 2000 and during its "restoration" in 2002. In any case the group has reanalyzed many data also coming from tests done in 1978 by the Shroud of TUrin Research Project (STURP) and has collected considerable information that clarifies the complex aspects of this sheet; hopefully this information will be improved upon when the Turin data become available.

In this document a list of facts directly related to the TS, subdivided into four sections, is presented. The first section describes unquestionable facts detected on the TS; the second one refers to confirmed observations or conclusions based on a proof made in reference to TS studies; the third one refers to facts or observations that were evidenced by some researcher but that are not universally accepted; the fourth one, assuming a scenario that the Shroud is actually the burial cloth of Jesus of Nazareth, includes correspondences with the Scriptures.

(1) Introduction

The TS is believed by many to be the burial cloth of Jesus of Nazareth when he was put in a tomb in Palestine about 2000 years ago. It has generated considerable controversy but unlike other controversial subjects (e.g., flying saucers and ghosts), the TS exists as a material object: it can directly and objectively be observed. The results of studies can be analyzed by scientific methods (Schwalbe 1982).

The TS is a linen sheet about 4.4 m long and 1.1 m wide, in which the complete front and back body images of a man are impressed. Of all religious relics it has generated the greatest interest. The cloth is handmade and each yarn (diameter about 0.25 mm) is composed of 70–120 linen fibers. Although not all scientists are unanimous, it has been shown by many scientists that the linen sheet enveloped or wrapped the corpse of a man who had been scourged, crowned with thorns, crucified with nails, and stabbed by a lance in the side. Also impressed are many other marks due to blood, fire, water, and folding, which have greatly damaged the double body image. Of greatest interest are the wounds, which, to forensic pathologists, appear to be unfakeable (Fanti and Moroni 2002).

The Shroud of Christ appeared in 1353 in Lirey, France, under mysterious circumstances and with no documentation whatever. In 1203, a soldier camping outside Constantinople with the Crusaders, who sacked the city the following year, noted that a church there exhibited every Friday the cloth in which Christ was buried, with the figure of his body. It is probable that this cloth and the TS are the same. It seems that the TS was among the spoils of the Crusades, together with many other relics brought back to Europe. Before the sacking of Constantinople in 1204 there are some documents that refer to the presence of the TS: for example, some characteristics of Christ reproduced in some Byzantine coins (gold *solidus*) of the seventh to thirteenth centuries A.D. are very similar to those of the TS body image.

I. Wilson (1998) identified the TS, folded four times to show only the face, with the *Mandylion*, a cloth said to have received the miraculous imprint of Christ's face and to have been taken to Edessa in the first century A.D. The tradition of this imprint "made without hands" developed first in the Byzantine Empire; a similar tradition arose in the seventh and eighth centuries in the West, that of Veronica, who wiped the brow of Christ with her veil and found an imprint of his face remaining.

Scientific interest in the TS developed after 1898, when S. Pia, who photographed it for the first time, noticed that the negative image on the TS looked like a photographic positive. Correlations with the anatomical characteristics of a human body were also very high and not comparable with anatomical characteristics normally depicted in popular Medieval art. In 1931, G. Enrie again photographed the TS at a very high resolution.

The TS has a front image 1.95 m long and a back image 2.02 m long, separated from the former by a nonimage zone of 0.18 m (measurements done before 2002); the images show an adult male, nude, well proportioned and muscular, and with a beard, mustache, and long hair.

The TS has been radiocarbon-dated to A.D. 1260–1390 (Damon et al. 1989), but a great number of scientists believe that the method used to take the sample and the reliability of radiocarbon dating are not satisfactory because the linen underwent many vicissitudes (e.g., fires, restorations, water, exposure to candle smoke, and the breath of visitors). For example, some researchers have proposed that the 1532 fire probably modified the quantity of radiocarbon in the TS, thus altering its dating, and others believe in the existence of a biological complex of fungi and bacteria covering the yarns of the TS in a patina (Moroni 1997, Garza-Valdes 2001). Recently it was demonstrated that the 1988 sample is not representative of the whole TS (Adler 1999, 2000, Marino 2000, 2002, Rogers 2002, 2005).

Many hypotheses and experimental tests have been carried out on linen fabrics to explain the formation of the body image, both in favor of authenticity and vice versa. Examples are:

(a) The body image is caused by the emanation of ammoniacal vapors (Vignon 1902).

(b) The body image is due to a chemical process similar to that which happens in leaves of herbaria: the image originated through direct contact (De Salvo 1982, Volckringer 1991).

(c) The body image is a painting (McCrone 1980).

(d) The body image is due to a natural chemical reaction (Rogers 2002).

(e) The body image was obtained from a warmed bas relief (Pesce Delfino 2001).

(f) The body image was obtained by rubbing a bas relief with pigments or acids (Nickell 1997).

(g) The body image was obtained by a modified carbon dust drawing transferred to the cloth by rubbing (Craig and Bresee 1994).

(h) The body image was obtained by exposing linen in a "darkened room" using chemical agents available in the Middle Ages (Allen 1998, Picknett and Prince 1994).

(i) The body image was obtained by exposing a linen cloth to sunlight with a glass plate containing an oil-painted image on its surface (Wilson 2005).

(j) The body image was obtained by surface electrostatic discharges caused by an electric field of seismic origin or directly generated by the enveloped Man (Scheuermann 1987, De Liso 2000, 2002, Lattarulo 2003, Fanti 2005, Fanti et al. Sept 2005).

(k) The body image is due to an energy source coming from the wrapped or enveloped Man, perhaps caused during the Resurrection (Lindner 2002, Rinaudo 1998, Jackson 1990, Moran 2002).

Although good experimental results have been obtained by a number of researchers, in the sense that, at first sight, the image, generally limited to the face, is similar to that of the TS Man, until now no experimental test has been able to reproduce all the characteristics found in the image impressed on the TS.

Some researchers interested in the TS scientific problems formed the Shroud Science Group on Yahoo! to discuss these issues via the Internet. A first objective posed by them is that regarding the possible explanation of the body image formation. To deepen the discussion in accordance with the scientific method, all the scientists agreed to define a list of evidence of the TS upon which to base their further debate. This paper, still in progress, presents the list of evidence defined by the researchers, which is intended to be useful for future discussion.

(2) List of Facts and Observations

The list is subdivided into four different types of evidence.

- **Type A** refers to unquestionable observations made on the TS numbered as "A*n*," where *n* is the evidence number.
- **Type B** refers to confirmed observations or conclusions based on a proof made in reference to TS studies and are numbered as "B*n*."
- **Type C** refers to facts or observations that were evidenced by some researchers but that are not universally accepted, and are numbered as "C*n*."
- Assuming a scenario that the TS is actually the burial cloth of Jesus of Nazareth, it makes sense to include the Scriptures in this discussion, not on a theological level, but describing some things that might have an impact on the TS; for this reason **Type D** refers to correspondences with those described in the holy texts and are numbered as "D*n*."

(2.1) Specific Facts

The list of **Type A** facts refers to unquestionable observations made on the TS and they are at the basis of every hypothesis formulation in the sense that an hypothesis must be tested against all type A facts and only if it is congruent with all of them, none excluded, can it be considered for further an in-depth study.

(2.1a) Chemical-Physical Characteristic of the Linen Yarns and Fibers

(A1) The yarn used to weave the Shroud was spun with a "Z twist" (Raes 1974, Vial 1989, Curto 1976, Pastore 1988).

(A2) Direct microscopy showed that the **image color** resides only on the **topmost fibers** at the highest parts of the weave (Evans 1978, Pellicori 1981).

(A3) Phase-contrast photomicrographs show that there is a very thin coating on the outside of all superficial linen fibers on Shroud samples, named **Ghost**; "Ghosts" are colored (carbohydrate) impurity layers pulled from a linen fiber by the adhesive of the sampling tape and they were found on background, light-scorch, and image sticky tapes (Zugibe and Rogers 1978, Rogers 2002).

(A4) Body image color resides on the **thin impurity layer** of outer surfaces of the fibers (Zugibe 1978, Heller 1981, Rogers 2002).

(A5) According to M. Evans (1978), photomicrographs ME-02, ME-08, ME-14, ME-16, ME-18, ME-20, ME-25, and ME-29, the color of the image areas has a discontinuous distribution along the yarn of the cloth: **striations** are evident. The image has a distinct preference for running along the individual fibers making up a yarn, coloring some but not others (Pellicori 1981, Schneider 2005). Fibers further from a flat surface, tangent to the fabric, are less colored, but a color concentration can be detected in correspondence to crevices where two or three yarns cross each other (ME-20) (Fanti 2005).

(A6) The cellulose of **the medullas** of the 10–20-micrometer-diameter fibers in image areas is **colorless** because the colored layer on image fibers can be stripped off, leaving colorless linen fibers (Heller 1981, Rogers 2002).

(A7) The **colored layers** in the adhesive have the same chemical properties as the **image** color on fibers (Rogers 2005).

(A8) The **crystal structure** of the cellulose of image fibers has **not visibly changed** with respect to that of the nonimage fibers (scorches have)(Rogers 2002, Feller 1994).

(A9) The colored coating cannot be dissolved, bleached, or changed by standard chemical agents, but it can be **de-colorized** by reduction **with diimide** (hydrazine/hydrogen peroxide in boiling pyridine); the residue from reduction is colorless linen fibers (Heller 1981, Rogers 2003).

(A10) The pyrolysis/mass spectrometry data showed the presence of **polysaccharides of lower stability** than cellulose on the surface of linen fibers from the TS (Rogers 2004).

(A11) Photomicrographs and samples show that the image is a result of **concentrations of yellow to light brown fibers** (Pellicori 1981, Jumper 1984, McCrone and Skirius 1980, Schwalbe 1982, Rogers 2002).

(A12) The image formation mechanism **did not char the blood** (Rogers 1978–1981).

(A13) The image formed at a relatively **low temperature** (Rogers 1978–1981).

(A14) The 1978 quantitative **X-ray fluorescence spectrometry** analysis detected significant uniform amounts of **calcium and strontium** concentrations (a normal impurity in calcium minerals) and iron in the Shroud (Morris 1980, Rogers 2003, Adler 1998).

(A15) Microchemical tests with iodine and pyrolysis/mass spectrometry detected the presence of **starch impurities** on the surfaces of linen fibers from the TS (Rogers 2002, 2004).

(A16) The **lignin** that can be seen at the wall thickenings and/or growth nodes of the linen fibers of the TS does **not** give the standard **test for vanillin** (Rogers 2002, 2005).

(A17) There are **no cementation** signs among the image fibers (Pellicori 1981).

(A18) **No painting pigments or media scorched** in image areas or were rendered water soluble at the time of the A.D. 1532 fire (Rogers 1977–1978, 1981/2002, Schwalbe 1982).

(A19) **No fluorescent pyrolysis products** were found in image areas (Rogers 2002).

(A20) After weaving, the TS yarns were **washed** with a very mild, natural material because of the presence of **flax wax** on the fibers and the specular reflectance of the nonimage fibers (Rogers 2003).

(2.1b) Optical Characteristics of the Cloth

(A21) The cloth shows **bands** of slightly different colors of yarn that are best observed in ultraviolet (UV) photographs. For example between face and hair there are two noncolored bands that continue along the warp direction (Miller and Pellicori, 1981, Fanti 2003, Rogers 2002, 2005).

(A22) There is a **correspondence** (even if not complete) between cloth **bands** of slightly different colors of yarn of the front and **back surface** (Ghiberti 2002, Fanti 2003).

(A23) The colored fibers in nonimage (background) areas show the **same type of superficial color** as body image fibers, their spectra are the same, and the cellulose in them is not colored (Gilbert 1980, Rogers 2002).

(A24) The **body image does not fluoresce** in the visible under UV illumination (Gilbert 1980, Pellicori 1981).

(A25) The nonimage area fluoresces with a maximum at about 435 nm (Pellicori 1981).

(A26) A **redder fluorescence** can be observed around the **burn holes** from the A.D. 1532 fire (Pellicori 1981).

(A27) The cloth does **not** show any **phosphorescence** (Rogers 2005).

(A28) All the chemical and microscopic **properties** of **dorsal and ventral** image fibers are **identical** (Jumper 1984).

(A29) An emission **image** was clearly **visible in the 8–14 micrometers infrared** (IR) range (Accetta 1980).

(A30) **IR emission** of the image at a uniform **room temperature**, and in the **3–5-micrometer range** was **below** the instrument **resolution** (Accetta 1980).

(2.1c) Body Image

(A31) The body image is **very faint**: reflected optical densities are typically less than 0.1 in the visible range (Jumper 1984, Schwalbe 1981).

(A32) The body image shows **no evidence of image saturation** (Jackson 1977, 1982, 1984).

(A33) The body image has a **resolution of 4.9±0.5 mm at the 5% MTF value** (e.g., the lips); the resolution of the bloodstains is at least 10 times better (e.g., the scratches in the scourge wounds) (Jackson 1982, 1984, Moran 2002, Rogers 2003, Fanti 2004-MTF, Fanti Sept. 2005-MTF).

(A34) The body image **does not have well-defined contours** (Jackson 1982, 1984, Moran 2002).

(A35) **A nonimage area** is detectable among the **fingers** of the TS image (Fanti 2004).

(A36) There is a **darker spot** in correspondence to the palm of the TS Man's hand near the index finger (Accetta 2001, Antonacci 2000).

(A37) The **thumbs** are not visible in the hand image (Bucklin 1982, Ricci 1989).

(A38) In correspondence to the **middle of the nose** there is a **swelling** (Fanti 2004).

(A39) Detailed photographs and microscopic studies of the cloth in the nose image area show **scratches and dirt** (Bucklin 1982).

(A40) The **hair** on the frontal image show **high luminance levels** relatively to the face: for example, the left hair is darker than the cheeks (Fanti 2004).

(A41) There is **no** evidence of **image between** the tops of the front and dorsal **heads** (Adler 1999, Moran 2002).

(A42) In the positive photograph of Durante (2000), the **luminance levels** of the front and back body images (face excluded) are **compatible** within an uncertainty of 5%; the front image is generally darker than the dorsal one (Moran 2002, Fanti 2005).

(A43) The image of the **dorsal side** of the body does **not penetrate the cloth** any more deeply than the image of the ventral side of the body (Jumper 1984, Rogers 2005).

(A44) **The luminance level of the head** image in the positive photograph of Durante (2000) is 10% and more lower (**darker**) than that of the whole-body image (Moran 2002).

(A45) The **image-forming mechanism** operated regardless of different body structures such as skin, hair, beard, and perhaps nail (Antonacci 2000).

(A46) The **thermograms** did not show the lower **jaw** of the image (Rogers 2003), even if it is visible (Whanger 1998).

(A47) A body image **color** is visible on the **back surface** of the cloth in the same position of some anatomic details as for the body image of the frontal surface of the TS. The **hair** appears more easily to the naked eye (Ghiberti 2002) but also other details of face and perhaps hands appear by image enhancement (Maggiolo 2002/03, Fanti and Maggiolo 2004).

(A48) **No image** color is visible on the **back surface** in correspondence to the **dorsal** image (Ghiberti 2002, Maggiolo 2002/03, Fanti and Maggiolo 2004).

(A49) The nose image on the back surface of the TS presents the same extension of both nostrils, unlike the **frontal**, in which the **right nostril** is less evident (Fanti and Maggiolo 2004).

(A50) Image details corresponding to the face **grooves** are more faintly represented (e.g., eye sockets and skin around the nose), convex **hills** on the face (e.g., eyeballs and nose tip), however, are more clearly represented (Scheuermann 1983).

(A51) Although anatomical details are generally in close agreement with standard human body measurements, some measurements made on the Shroud image, such as **hands, calves, and torso, do not agree with anthropological standards** (Ercoline 1982, Simionato 1998/99, Fanti and Faraon 2000, Fanti and Marinelli 2001).

(A52) The body image shows **no** evidence of **putrefaction** signs, in particular around the lips. There is no evidence of tissue breakdown (formation of liquid decomposition products of a body) (Bucklin 1982, Moran 2002).

(A53) **No image is formed under the bloodstains** (Heller 1981, Schwalbe 1982, Brillante 2002).

(A54) The front image shows **hair** that **goes down to the shoulders** (Fanti and Faraon 2000).

(A55) The image of the TS Man appears as if he was **scourged** (Bucklin 1982, Ricci 1989).

(A56) The image of the TS Man appears as if he was **crucified**: it appears with nail holes and corresponding blood at the wrists and top of the feet (Bucklin 1982, Ricci 1989).

(A57) The image of the TS Man demonstrates no evidence of **maiming** or disfigurement (Bucklin 1982, Ricci 1989).

(2.1d) Blood and Body Fluids

(A58) **Body fluids** other blood or serum than did not percolate into the cloth (Rogers 2003).

(A59) The blood or serum has migrated by **capillary inhibitions** from the warp side to the weft side of the TS Man and, depending on their abundance and consistency, they filled the mesh apertures (Fanti 2004).

(A60) There is a class of particles on the TS ranging in color from red to orange that test as blood-derived residues. They test positive for the presence of protein, hemin, **bilirubin**, and albumin, give positive hemochromagen and cyanmethemoglobin responses, and after chemical generation display the characteristic fluorescence of **porphyrins** (Adler 1999).

(A61) The blood on the TS is **not denatured**. Therefore both the image formation mechanism and the 1532 fire did not involve processes that would denature the blood (Rogers 2004).

(A62) The **blood** from the large flow on the back **darkened** (scorched) at an adjoining **scorch** (Rogers 1978).

(A63) The **red flecks** McCrone (1980, 2000) claimed were hematite had an **organic matrix** (Heller 1983, Rogers 2004).

(A64) Microscopic observation of **blood flecks** of sample 3EB showed specular reflection: **the blood went onto the surface as a liquid** (Rogers 1978).

(A65) **Blood** spots are much **more visible on the TS by transmitted light** than by reflected light; this implies that the blood saturated the cloth and it is not a superficial image as the body image is (Rogers 1978).

(A66) **Many blood** traces visible on the frontal image are also visible **on the back** image in the **same position** (Fanti 2003).

(A67) **Bloodstains** are well marked on the **reverse side**, although they are fainter than on the front side of TS (Fanti 2003, Whanger 2004).

(A68) Some human **bloodstains** appear on and **outside of the body image** (left elbow) (Heller 1980, 1981, Baima Bollone 1981, 1982, Jackson, 1987, Carreira, 1998).

(A69) In correspondence to the **knees** on the dorsal image, there are scourge marks in correspondence to **lower luminance levels** of the body image (Fanti 2003).

(A70) The **blood** on the TS **does not fluoresce** in UV illumination (no porphyria and no fluorescent pigments) (Rogers 1978).

(A71) The **blood** on the TS can be removed with a **proteolytic enzyme** (Adler 1999).

(A72) **No smears** are evident in the blood traces (Bucklin 1982, Ricci 1989, Antonacci 2000).

(A73) **No potassium** signals could be found in any of the blood area data (Morris 1980).

(A74) In UV fluorescence the **scourge** marks appear with **dumb-bell shapes** (Bucklin 1982, Ricci 1989).

(A75) In UV fluorescence the **scourges** are resolved into **fine scratches**: three and, in some cases, four parallel scratches can be distinguished (Bucklin 1982, Ricci 1989).

(A76) The bloodstain corresponding to the right side of the **chest**, sixth rib, shows **separation of blood** from a clearer liquid material (Bucklin 1982).

(A77) The **DNA** found in blood spots is badly **degraded** (Rogers 2005).

(A78) **No broken fibers** were found under the blood clots (Svensson 2005).

(2.1e) Others

(A79) **Earthy material** (limestone composed of aragonite with strontium and iron) was found on the **feet** of the TS Man (Kohlbeck 1986, Nitowski 1986, 1998, Antonacci 2000). Earthy material was also found in correspondence to the **nose** and the **left knee** (Pellicori 1981).

(A80) Drops of **wax** were found (Maloney 1989).

(A81) Microscopic observation of the bridge of the **nose** showed discontinuous distribution of **light gold-colored fibers**. All were on the top of the yarn (Rogers 1978, 2004).

(A82) There is **no** observed microscopic, chemical, or spectroscopic evidence for the presence of any **dry powder** responsible for the body image on the TS (Adler 1999).

(A83) Some little **black spots** (diameter of 1–2 mm) appear out of the body image (e.g., near the head, between the hair, and the water stain); they are also visible, in the same position, on the back surface of the TS (Maggiolo 2002/03, Fanti 2003, Rogers 2003).

(A84) Large **water stains** are visible on both sides of the cloth (Fanti 2004).

(A85) **Silver** traces were found around the burn holes in the scorch area of the TS (Heller 1983).

(A86) The white cloth used to cover the display board for the showing (1978) was fluorescent. Rudy Dichtel reported many intensely **fluorescent short fibers on the surface of the Shroud** (Rogers 2004).

(A87) **Aldehyde and carboxylic acid functional groups were detected** in the TS fibers (Adler 1981).

(2.2) Confirmed Observations

Type B refers to confirmed observations or conclusions based on a proof made in reference to the TS. Therefore these observations must also be used to test any new hypothesis.

(2.2a) Chemical-Physical Characteristic of the Linen Yarns and Fibers

(B1) The TS samples examined have **herringbone 3:1** twill weave (Vial 1989).

(B2) **Traditional dimensions** of the TS of 436 × 110 cm (Baima Bollone 1978) are changed after 2002 "restoration": one side (the lower considering horizontal the body image, with the frontal side on the left) measured 437.7 cm in 2000 and 441.5 cm in 2002; the opposite side measured 434.5 in 2000 and 442.5 in 2002; its height of 112.5 and 113 cm, respectively, on the left and on the right in 2000 but 113.0 and 113.7 cm in 2002 (Ghiberti 2002). A measurement made in 1868 by Gastaldi (Baima Bollone 1978) reports the following dimensions: 410 × 140 cm (Scarpelli 1983).

(B3) The **thickness** of the cloth measured by Jackson with a micrometer is variable from 318 to 391 micrometers (Rogers 2004).

(B4) There appears to be more variation in the **diameter** of warp yarns than weft Rogers (1978).

(B5) The TS weave is **very tight** (Raes 1974, Rogers 1978, Vial 1989).

(B6) Although yarns and design of **Raes samples** look like the main part of the cloth, linen fibers from the Raes samples that was cut in 1973 are chemically **different** (from reflected spectroscopy and chemical analysis) (Adler 2000, Rogers 2002).

(B7) **Cotton** fibers were found in the Raes samples and they were identified as *Gossypium herbaceum*, a common Middle East variety (Raes 1974, 1991).

(B8) The **sewing** connecting the upper linen **band** of the TS is very particular and typical of very old manufacture (Flury Lemberg 2000, 2001).

(B9) Reflectance spectra, chemical tests, laser-microprobe Raman spectra, pyrolysis mass spectrometry, and X-ray fluorescence all show that the image is **not painted** with any of the expected, historically documented pigments (Schwalbe 1982, Morris 1980, Heller 1981, Mottern 1979).

(B10) Chemical tests showed that there is **no protein painting medium** or protein-containing coating in image areas (Rogers 1978–1981, Heller 1981, Pellicori 1980, 1981, Gilbert 1980, Accetta 1980, Miller 1981).

(B11) The image fibers do not show any sign of **capillary** flow of a colored or reactive liquid (Evans 1978, Pellicori 1981).

(B12) Flakes of image color can be seen in other places where they fell off and stuck to the adhesive. The chemical **properties of the coatings** are the **same** as the image color on image fibers. All of the color is on the surfaces of the fibers (Rogers 2002, Heller 1981).

(B13) There are **no pigments** on the body image in a sufficient quantity to explain the presence of an image (Pellicori 1981).

(2.2b) Optical Characteristics of the Cloth

(B14) The TS linen has a **lustrous** finish (Rogers, 1978–1981).

(B15) If a fiber is colored, it is **uniformly colored** around its cylindrical surface (Adler 1996, 1999); relatively long fibers show variation in color from nonimage to image area (Fanti 2004).

(B16) A **crease** below the **chin** of the image: on the frontal surface of the TS, the inside part of crease has a lighter color similar to the background, but it has darker margins similar to the image color. On the back of the cloth, the same crease is darker in correspondence to the lighter color of the frontal surface and the margins are confused with the background: the darker margins are of the same straw-yellow color of the body image (Rogers 2004).

(B17) In the **UV** emission and absorption photographs the **background** cloth shows a light **greenish yellow** emission (Adler 2002).

(B18) Where one of the **image yarn crosses over another**, there is often **no color** on the lower one (Heller 1983, Rogers 2005).

(B19) The image of the **dorsal side** of the body shows fairly the **same color** density and distribution as the ventral (Jumper 1984).

(B20) The **IR photograph** of the face made by Judica Cordiglia, if compared with visible photographs of the face, indicates the **low absorption near the IR** of the products of image formation (Cordiglia 1974, Accetta 1980, Rogers 2003).

(2.2c) Body Image

(B21) Up to now, all the **attempts to reproduce a copy of the TS** similar in all the detected characteristics have **failed** (Carreira 1998, Fanti 2004).

(B22) The **most of the prominent parts** in the vertical direction (nose, beard, sole, calf) of the body image **are marked** (Fanti 2003).

(B23) The **hair** on the front image is **soft** and not matted, as would be expected if it were soaked with a liquid (Fanti 2004).

(B24) When their lengths are measured, the **dorsal image is longer** than the ventral image in a manner similar to the imprint on a sheet of a man having the head tilted forward, his knees slightly bent, and his feet extended (Craig 2003, Cagnazzo 1997/98, Fanti 2000).

(B25) The **frontal body image** (195 cm long) is **compatible**, within an uncertainty of ± 2 cm, with the **dorsal** image (202 m long) if it is supposed that the TS enveloped a corpse having the head tilted forward, the knees partially bent, and the feet stretched forward and downward (Basso 2000).

(B26) On the basis of cloth measurements (Baima Bollone 1978), the image corresponds to **a man 175±2 cm tall** (Simionato 1998/99, Faraon 1998/99, Basso 2000).

(B27) The body image has the **normal tones of light and dark reversed** with respect to a photograph, such that parts nearer to the cloth are darker (Jumper 1984, Craig 2004, Schneider 2004).

(B28) The luminance distribution of both the frontal and dorsal images has been correlated to the clearances between a **three-dimensional** surface of the body and a covering cloth (Quidor 1913, Sullivan 1973, Gastineau 1974, Jackson 1977, 1982, 1984, Fanti 2001, Moran 2002).

(B29) The luminance distribution of the body image can be correlated with a **highly directional mapping function** (Jackson 1977, 1982, 1984).

(B30) The body image shows **nondirectional light sources** in the sense that there are no shadows, cast shadows, highlights, and reflected lights in or on the body image (Moran 2002, Craig 2003).

(B31) The absence of saturation implies that the **image formation did not "go to completion,"** that is, it did not produce the maximum number of conjugated carbon–carbon double bonds (Rogers 2003, Gilbert a1980: Figs. 8 and 10).

(B32) In correspondence to image sections of cylindrical elements such as legs, the **luminance level variation** approximates a **sinusoidal law** (Fanti 2004).

(B33) In reference to a cloth wrapping a body, there is **no** evidence of body **image** formation at the **sides** of the body on both the frontal and dorsal TS images (Adler 1999, Moran 2002).

(B34) The **Fourier transform** of the body image shows a nearly continuous spectrum in correspondence to the spatial frequencies up to 100 [1/m] (Fanti 1999, Maggiolo 2002/03).

(B35) The body image indicates the **absence of brush strokes** (Lorre 1977).

(B36) The frontal image, at least in correspondence to the head, is **doubly superficial** (Fanti and Maggiolo 2004).

(B37) The **fingers** in the image appear to be **longer** than average for a man, but they are still within the normal range (Gaussian distribution) (Heller 1983, Whanger 2005).

(B38) **Image distortions** of hands, calves, and torso on the TS of are very close to those obtained by a man enveloped in a sheet (Ercoline 1982, Simionato 1998/99, Fanti and Faraon 2000, Fanti 2001).

(B39) The very **high rigidity of the body** is evident on the back image, especially in correspondence to the gluteal: the anatomical contours of the back image demonstrate minimal surface flattening (Bucklin 1982, Basso 2000).

(B40) The image of the TS Man shows the effects (wounds) of many **pointed objects** (Bucklin 1982, Ricci 1989).

(B41) The **tibiofemoral anthropometric index** of the image of the TS Man is 83% (Fanti 1999).

(B42) **No broken bones** are evident on the body image (Bucklin 1982, Ricci 1989).

(B43) There is a **swelling** on the **face** over the right cheek (Bucklin 1982).

(B44) There is a slight **deviation of the nose** and at the tip of the nose is an area of discoloration (Bucklin 1982).

(B45) A **body image** is visible in areas of body–sheet **noncontact zones**, such as those between nose and cheek (Fanti 2004).

(2.2d) Blood and Body Fluids

(B46) There is a first **type of bloodstain** that corresponds to the blood **exudated from clotted wounds** and transferred to the cloth by being in contact with a wounded human body such as scourging and a crown of thorns wounds or wrist wounds (Adler 1999).

(B47) There is a second **type of bloodstain** that corresponds to the blood that directly **flowed** on the TS, such as feet wounds or a side wound with blood separation in a dense and a serous portion (Brillante 2002, Schneider 2004).

(B48) The UV photographs of single bloodstains show a distinct **serum clot retraction ring** (Adler 1999).

(B49) The chemical and physical parameters of the **bloodstains are different** than mineral compositions proposed by **artists** (Adler 1999).

(B50) The bloodstains observable on the back surface have been described as **imbibed flows** throughout the cloth (Ghiberti 2002).

(B51) Blood traces on the back surface of the TS are smaller in size when compared with the corresponding traces on the frontal side, showing that **blood was transposed** onto the cloth **touching the frontal** side of the TS (Fanti 2003).

(B52) The maintenance of the **bright red color** of the TS **blood** with time was observed, but the explanation of why the color is so red is not definitive (Brillante 2002).

(B53) There are **blood traces not consistent** with scalp hair traces soaked with blood in correspondence to the image of the **hair** on the front side (Lavoie 1983, Fanti 1999).

(B54) The **wrist wound position** can be referred to as the hand nail used for the crucifixion (Fanti and Marinelli 2003).

(B55) The **blood clots** were transposed to the linen fabric during **fibrinolysis** (Brillante 1983, Lavoie 1983). The process of fibrinolysis could cause clots to liquefy sufficiently for the blood to transfer to the cloth as a serous-laden liquid rather than a moist jelly-like substance (Craig 2004).

(B56) Some bloodstains are comparable to transfers that would be expected if the arms were posed in a **nonhorizontal position** (Lavoie 1983, 2003, Fanti 2005, Schwortz 2005).

(2.2e) Others

(B57) The **limestone** found on the feet contains calcium in the form of **aragonite**. Similar characteristics were found on samples coming from Ecole Biblique tomb in Jerusalem (Levi-Setti 1985, Antonacci 2000).

(B58) It is **unknown** whether *Saponaria officinalis* can be detected on the Shroud (Rogers 2003, Jumper 1984, Gilbert 1980).

(B59) **Rust stains** due to thumb tacks were found on the sides of the TS (Faraon 1998/99, Schwortz 2003).

(B60) Characteristics of the TS face and right foot are close to those found on some **Byzantine coins** (gold *solidus*) of the seventh to thirteenth centuries A.D. (Moroni 1986).

(B61) Some **water stains** are older than the 1532 fire, because they indicate a **different folding** of the TS (Guerreschi and Salcito 2002).

(2.3) Evidence to be Confirmed

Seeing things and not seeing things are perhaps the biggest problems in legitimate Shroud research. "I think I see" and "I don't see" seem to be the underpinning of many "scientific" analyses. The body image on the Shroud was formed by some process. We don't know, for now, what that was, nor the shape of the cloth, nor the environment where the body was positioned: we can only suppose what that might have been; we don't know many variables.

Our brain–eye system may play tricks on the researcher. Because of a priori assumptions, it may be that he or she perceives things that conform to something searched for, and conversely, he may fail to perceive images because of not knowing what various objects look like. Many of the images are below the ordinary human perceptual threshold; therefore anything must be probed for documentable facts, including using image enhancement techniques.

Type C refers to facts that were evidenced by some researchers but that are not universally accepted; therefore they can help in formulating new hypotheses, but they cannot be used to test a new hypothesis.

(2.3a) Body Image

(C1) The chiaroscuro effect is caused by a different number of yellowed fibers per unit of surface so that this is an image with **"areal" density** (Moran 2002, Fanti and Marinelli 2003).

(C2) Body image characteristics can be referred to the hypothesized effect of a man becoming **mechanically transparent** that radiated a burst of energy (Jackson 1977, 1984, 1990).

(C3) The TS face shows a sad but majestic **serenity** (Moroni 1997).

(2.3b) Others

(C4) **Pollen grains** relative to the zones of Palestine, Edessa, Constantinople, and Europe were found (Frei 1979, 1983, Danin 1999).

(C5) **Pollen grains** with **incrustations** soluble in water were found from the vacuumed samples taken from the back surface of the cloth (Riggi 2003).

(C6) The wrapped or enveloped body was a **corpse** (Bucklin 1982, Lavoie 1983, Jackson 1998, Petrosillo 1988, Brillante 2002, Baima Bollone 2000, Fanti 2003, Zugibe 2005), but someone still states that the body was in a state of **coma** (Bonte 1992, Hoare 1994, Gruber 1998, Kuhnke 2004, Felzmann 2005).

(C7) The human blood is of **AB group** (Baima Bollone 1981, 1982).

(C8) The **radiocarbon dating** of 1988 states that the TS linen has an age of 1260–1390 (Damon et al. 1989).

(C9) "Preliminary estimates of the kinetics constants for the loss of **vanillin** from lignin indicate a much older age for the cloth than the radiocarbon analyses" (Rogers 2005).

(C10) There is the image of an identified **coin** (dilepton lituus) on the right eye (Filas 1982, Haralick 1983, Barbesino 1997).

(C11) There is an image of another identifiable **coin** (Pilate lepton simpulum) over the **left eye** (Balossino 1997, Barbesino 1997).

(C12) The TS is like a **funerary sheet** (Persili 1998).

(C13) There are some **analogies** between the TS and the **Oviedo Sudarium**, including many congruent bloodstains (Whanger 1996).

(C14) There are various **writings** around the face (Marion 1998).

(C15) There are many identified **floral images** on the TS, which indicate that the Shroud originated in the vicinity of Jerusalem in the spring of the year and which have the appearance expected from corona discharge. Some images are consistent with the **fruits** of *Pistacia* plants, which were used as burial spices (Danin, 1999, Whanger, 2000).

(C16) **Human DNA** comes from Riggi's blood samples from the TS—this because three gene segments were cloned and studied (Garza-Valdes 2001).

(C17) Results from the DNA analysis, made from the TS blood at the University of Texas, San Antonio, USA, indicate that some genetic characteristics are relative to the **Semitic** race (e.g., hair) (Riggi 2003).

(C18) The TS Man died because of an **infarct** followed by **hemopericardium** (Malantrucco 1992).

(C19) Some **teeth** are visible on the image (Whanger 2000, Accetta 2001).

(C20) The **skull** is visible on the TS (Whanger 2000).

(C21) Images of the **bones** of the fingers, of the palms (metacarpals), and of the wrist are visible, and in particular a hidden thumb (Whanger 2000, Accetta 2001).

(C22) A **sponge** is visible on the TS (Whanger 2000).

(C23) A large **nail** with two crossed smaller nails are visible on the TS (Whanger 2000).

(C24) A **shaft** and head of spear are visible on the TS (Whanger 2000).

(C25) A **crown** of thorns with stalks and flowers is visible on the TS (Whanger 2000).

(C26) Some **bloodstains** such as those on the arms and the "reverse 3" on the forehead present a **discontinuity** in which a more attenuate region is evident(Jackson 1987, Schneider 2004).

(C27) Several wood tubules were found from an **oak** from Riggi's samples (Garza-Valdes 2001).

(C28) A **bioplastic coating** was found around the TS linen fibers (Garza-Valdes 2001).

(C29) Traces of **saliva** are visible on the image (Scheuermann 1983).

(C30) Traces of **tears** may be visible on the body image under the right eye (Guerreschi 2000).

(C31) An **ecchymosis**, on the left shoulder blade level, and a wound on the right shoulder that added to the wounds of the scourge are evident; in such areas the wounds caused by the scourge appear enlarged probably by the pressure of the **patibulum** (Ricci 1989).

(C32) Some **early paintings** of Jesus (before the sixth century A.D.) in Rome have been produced independently from the TS but have a significant similarity to the image on the TS. If it is assumed that these paintings go back to people who have known Jesus personally and knew therefore how he looked like. The significant similarities to the image on the TS indicate that both types of images go back to the same source: the historical Jesus (Felzmann 2003–2005).

(C33) **Natron** (sodium carbonate) was found in the dusts aspired from the back surface of the TS (Riggi 1982).

(C34) **Aloe and myrrh** were found by microscopic analysis (Baima Bollone 1983 and Nitowski 1986) but not by Heller (1983) and Rogers 2003).

(C35) The **scourge marks are part of the image** and primarily not caused from blood coming out of the wounds (Hoare 1994).

(C36) A **ponytail** is visible on the back image (Fanti and Marinelli 2001, Fig. 12 B and C, Antonacci 2000, Fig. 3).

(C37) In the image of **the back of the head** some **bloodstains** are partially **masked** (Scheuermann 1984).

(C38) Some bloodstains are comparable to transfers that would be expected if a person was posed in the **vertical position** (Lavoie 1983, 2003).

(2.4) Analogies between the TS Man and Christ, from the Old and the New Testament

It is hypothesized by many researchers that the TS is the burial cloth of Jesus of Nazareth. The following list presents passages in the Scriptures that have an impact on the TS. If these Scriptures are accepted as a historic document, **type D** facts can be useful to verify the proposed hypotheses.

(D1) *And no sign shall be given to it except the sign of the prophet **Jonah** (Matthew 16:4). And all flesh shall see the salvation of God (Luke 3:6). And I am with you always, even to the end of the age (Matthew 28:20).* The TS shows a sign promised by

Jesus: like Jonah "who remained for three days in the stomach of the big fish," the Man of the TS remained for three days inside the sepulcher (Rodante 1987).

(D2) *A woman came to him with an alabaster jar of very expensive* **perfume**, *which she poured on his head as he was reclining at the table* (Matthew 26:7); *When she poured this perfume on my body, she did it to prepare Me for burial* (Matthew 26:12). Less than 48 hours before his crucifixion, the hair of Jesus was anointed with a very valuable oil and this fact must be considered for an hypothesis about the TS image formation (Scheuermann 1984).

(D3) *Then Pilate took Jesus and* **scourged** *Him* (John 19:1). *I offered my back to those who beat Me, my cheeks to those who pulled out my beard; I did not hide my face from mocking and spitting* (Isa 50:6). The whole body of the TS Man is cruelly scourged, except for the breast, where, hitting, one could cause death. The scourging was given like punishment apart, more abundant (120 strokes), from the normal (39 strokes) as a prelude to crucifixion (Zaninotto 1984).

(D4) *Then they* **struck** *Him on the head with a reed and spat on Him* (Mark 15:19). *And they struck Him with their hands* (John 19:3). The TS Man was hit on his face: for instance, various tumefactions and the breakage of the nasal septum are evident (Fanti and Marinelli 1998).

(D5) *And the soldiers twisted a* **crown of thorns** *and put it on His head* (John 19:2). *When they had twisted a crown of thorns, they put it on His head* (Matthew 27:29, etc.). The TS Man was crowned with thorns. The head presents many wounds caused by sharp bodies (Fanti and Marinelli 1998).

(D6) *And He,* **bearing His cross**, *went out to . . . (the) Golgotha* (John 19:17). The TS Man presents on the shoulders excoriations imputable to the transport of the horizontal part of the cross (patibulum) (Ricci 1989).

(D7) *Now as they came out, they found a man of* **Cyrene**, *Simon by name. Him they compelled to bear His cross* (Matthew 27:32). The TS Man fell repeatedly to the ground; this is demonstrated by the dust particles on the nose and on the left knee. Likely he was helped in the transport of the cross (Fanti and Marinelli 1998).

(D8) *My throat is **dry*** (Psa 69:3), *And for my thirst they gave me vinegar to drink* (Psa 69:21). From the forensic medicine analysis it results that the TS Man died dehydrated (Intrigillo 1998).

(D9) *Where (on the Golgotha) they **crucified** Him* (John 19:17). *They pierced My hands and My feet. I can count all My bones* (Psa 22:16–17). *You have taken by lawless hands, have crucified, and put to death* (Act 2:23). The TS Man, too, was crucified (Fanti and Marinelli 1998).

(D10) *Reproach has **broken my heart*** (Psa 69,20). *And Jesus **cried out** again with a loud voice, and yielded up His spirit* (Matthew 27:50). *Because for Your sake I have borne reproach; Shame has covered my face* (Psa 69:8). *My heart is like wax; It has melted within Me* (Psa 22:14). The hemopericardium, diagnosed in the TS Man as a consequence of the infarct, causes a violent dilatation of the pericardic pleura with consequent shooting pain from the back breast bone and immediate death (Malantrucco 1992).

(D11) *And saw that He was already dead, they **did not break His legs*** (John 19:33). *Nor shall you break one of its bones* (Exo 12,46). Contrary to many Roman crucifixions, they did not break the TS Man legs (Fanti and Marinelli 1998).

(D12) *But one of the soldiers **pierced His side** with a spear* (John 19:34), *But He was wounded for our transgressions* (Isa 53,5). *Then they will look on Me whom they pierced* (Zec 12:10). The TS Man too was pierced in the side after his death (Zaninotto 1989).

(D13) *And immediately **blood and water came out*** (John 19:34). *Flowing from under the threshold of the temple toward the east, for the front of the temple faced east* (Eze 47:1). *This is He who came by water and blood Jesus Christ; not only by water, but by water and blood* (1John 5:6). The TS Man also presents a blood and serum flow (Malantrucco 1992).

(D14) *And Nicodemus, who at first came to Jesus by night, also came, bringing a **mixture of myrrh and aloes**, about a hundred pounds* (John 19:39). *Then they took the body of Jesus, and* *bound it in strips of linen with the spices, as the custom of the Jews is to bury* (John 19:40). Some researchers state that the TS body was buried with aromatics such as aloe and myrrh because they found their traces on the cloth (Baima Bollone 1983).

(D15) *When Joseph had taken the body, he **wrapped it in a clean linen cloth** (or shroud), and laid it in his new tomb* (Matthew 27:59–60). The TS Man, too, was enveloped or wrapped in a new and expensive sheet bought by a wealthy person (Fanti and Marinelli 1998).

(D16) ***Nor will You allow** Your Holy One to see **corruption*** (Act 2:27). *For You will not leave my soul in Sheol, Nor will You allow Your Holy One to see corruption* (Psa 16:10). The TS does not show signs of putrefaction (Fanti and Marinelli 1998).

(D17) *You shall let **none of it (the Lamb) remain** until morning, and what remains of it until morning you shall **burn with fire**. It is the Lord's Passover* (Exo 12:10). Some researcher states that the TS presents a double sign: the disappearance and the burning, if one refers to the radiant hypothesis (Rinaudo 1998).

(D18) *For as **lightning** that comes from the east is visible even in the west, so will be the coming of the Son of Man* (Matthew 24,27); *For the Son of Man in his day will be like the lightning, which flashes and lights up the sky from one end to the other* (Luke 17,24). The German theologian G. Schwarz (1986) rectifies the Bible by translation into the Aramaic language. It seems, in doing so, he found the Shroud image–forming process in the Bible independent on the TS: "As a flash in lightning and shining: so I will exist in my day!" ("my day" = the day of Jesus Resurrection) (Scheuermann 1987).

(D19) *There was a violent **earthquake**, for an angel of the Lord came down from heaven and, going to the tomb, rolled back the stone and sat on it* (Matthew 28,2). Someone hypothesizes the presence of an earthquake as a cause of the body image formation (Judica Cordiglia 1986, DeLiso 2002, Lattarulo 2003).

(D20) *Then the other disciple, who came to the tomb first, went in also; and **he saw and believed**. For as yet they did not* *know the Scripture, that He must rise again from the dead* (John 20:8-9). *David, foreseeing this, spoke concerning the resurrection of the Christ* (Act 2:31). One hypothesis states that the TS Man became mechanically transparent with respect to the sheet and shed a flash of energy that would be the cause of the body image formation (Jackson 1990). Perhaps the particular shape of the TS seen by John induced him to believe in Christ's Resurrection.

(D21) *After that,* **He appeared to more than five hundred of the brothers** *at the same time, most of whom are still living, though some have fallen asleep* (1. Cor. 15:6). Paul has written this letter in the year 53–55. The time is too short (eyewitnesses) that all this might be an invention without historical nucleus (Felzmann 2003).

(3) Conclusions

The first goal posed by the researches of the Shroud Science Group on Yahoo! in order to better understand the TS has been reached: a list of evidence of the TS upon which to base their further debate on the body image formation problem has been defined, even if the work, is still in progress. Obviously some open questions will be easier to solve if the Turin officials become open to sharing new results and those obtained in 2002 to the Shroud Science Group and to any credible researcher interested the study about the most important relic of Christianity.

In consideration of space limitations, the facts have been stated in very simplistic terms, but the rich bibliography enclosed will allow the reader to go far more in depth in reference to the argument of interest.

Many hypotheses have been presented and some natural hypotheses are under test, but hypotheses involving the Resurrection of Jesus of Nazareth cannot be rejected. Among them there are hypotheses correlated to an energy source coming from the enveloped or wrapped Man, others correlated to surface electrostatic discharges caused by an electric field, or others correlated to natural chemical reactions also helped by the body fluids transferred to the cloth, but none, scientifically testable, simultaneously satisfies all the facts detected on the Shroud here reported. On the other hand other hypotheses such as that of Jackson (1990) seems to satisfy almost all the presented facts, but it is not scientifically testable because it bases itself on a nonscientific fact: the mechanically transparent Man.

The next goal of the Shroud Science Group on Yahoo! will be the presentation of all the possible hypotheses about the body image formation in a detailed form in order to test them against the facts reported in this paper. Each hypothesis should have a title in reference to the technique involved in the TS formation, not just considering the body image formation; the author's name who first proposed the hypothesis coupled with the researcher's name who presents it; a detailed technique description for the formation of both the body image and the bloodstains; possible correlation or interferences with the formation of other stains such as water; and comments and bibliographic references.

A.3 Cap. 3: Probability Calculation

We can sketch a quick calculation of the probabilities considering the 12 features observed in Fig. 3.33; to each of them an appropriate occurrence probability is assigned, assuming that there is no direct observation by the artist of the particular in a reference image but that the artist arbitrarily decides to reproduce that particular feature without being influenced by any reference. The probabilities assigned to each of the 12 features are the following:

(1) Hair: 1 chance in 100 ($P_1 = 0.01$)
(2) Arched eyebrow: 1 chance in 50 ($P_2 = 0.02$)
(3) "Reversed 3": 1 chance in 100 ($P_3 = 0.01$)
(4) Eyes: 1 chance in 10 ($P_4 = 0.10$)
(5) Right eye's contusion: 1 chance in 20 ($P_5 = 0.05$)
(6) Swollen cheekbones: 1 chance in 50 ($P_6 = 0.02$)
(7) Nose: 1 chance in 50 ($P_7 = 0.02$)
(8) Mustache: 1 chance in 50 ($P_8 = 0.02$)
(9) Sparse beard on the right side: 1 chance in 100 ($P_9 = 0.01$)
(10) Beard's shape: 1 chance in 50 ($P_{10} = 0.02$)

(11) Beard's gap below the lower lip: 1 chance in 5 ($P_{11} = 0.20$)

(12) Shape of the wrinkle on the neck: 1 chance in 10 ($P_4 = 0.10$)

Let us consider these two events in the probabilistic calculation:

- Event T (truth): the artist reproduces the face of the coin in Fig. 3.34 by chance.
- Event F (false): the artist is not able to reproduce these details by chance, but he or she needs a reference image.

The two events are mutually exclusive and both have a marginal probability P^I of 50%, so $P^I(T) = 0.50$ and $P^I(F) = 0.50$. Thus, by applying Bayes formula:

$$P^{III}(T) = \frac{P^I(T) \cdot P^{II}(T)}{P(E)},$$

in which the probability *P(E)*, that is, precisely the Bayes constant, is:

$$P(E) = P^I(T) \cdot P^{II}(T) + P^I(F) \cdot P^{II}(F).$$

In fact, since the two events T and F exclude each other, for the principle of total probability the mixed possibilities must be ruled out.

Under the hypothesis of two mutually excluding events and with these marginal probabilities, the posterior probabilities $P_i^{II}(T)$, $P_i^{II}(F)$ are evaluated for each of the 12 characteristics. Experimental verification would be required to have values with smaller uncertainty, but since in this case what matters is the magnitude of the result, we accepted the following assignments that may also have uncertainties of 0.02 in magnitude or perhaps even more in some cases, as P_8^{II}. The assignments are reported in Table A.1.

The posterior combined probabilities $P^{II}(T)$ and $P^{II}(F)$ are:

$$P^{II}(T) = \prod_{i=1}^{12} P_i^{II}(T) = 3.2 \cdot 10^{-18} \quad P^{II}(F) = \prod_{i=1}^{12} P_i^{II}(F) = 0.44$$

The final probabilities *P** without the Bayes constant *P(E)* are:

$$P^*(T) = P^I(T) \cdot P^{II}(T) = 0.50 \cdot 3.2 \cdot 10^{-18} = 1.6 \cdot 10^{-18}$$

$$P^*(F) = P^I(F) \cdot P^{II}(F) = 0.50 \cdot 0.44 = 0.22$$

Table A.1 Probabilities assigned to each of the 12 features of Fig. 3.34 for the event true (T) and false (F)

$P_1^{II}(T) = 0.99$	$P_1^{II}(F) = 0.01$	$P_7^{II}(T) = 0.98$	$P_7^{II}(F) = 0.02$
$P_2^{II}(T) = 0.98$	$P_2^{II}(F) = 0.02$	$P_8^{II}(T) = 0.80$	$P_8^{II}(F) = 0.20$
$P_3^{II}(T) = 0.99$	$P_3^{II}(F) = 0.01$	$P_9^{II}(T) = 0.99$	$P_9^{II}(F) = 0.01$
$P_4^{II}(T) = 0.90$	$P_4^{II}(F) = 0.10$	$P_10^{II}(T) = 0.98$	$P_10^{II}(F) = 0.02$
$P_5^{II}(T) = 0.95$	$P_5^{II}(F) = 0.05$	$P_11^{II}(T) = 0.80$	$P_11^{II}(F) = 0.20$
$P_6^{II}(T) = 0.98$	$P_6^{II}(F) = 0.02$	$P_12^{II}(T) = 0.90$	$P_12^{II}(F) = 0.10$

Then, to calculate the final probability of the true event, $P^{III}(T)$, the previously indicated Bayes formula must be used, in which $P(E)$ = 0.2200000000000000016. So the value calculated is:

$$P^{III}(T) = 7.26 \cdot 10^{-18}.$$

Thus the probability that the artist has fortuitously got that particular result are seven chances in one billion of billions.

A.4 Cap. 5: The Cellulose

Cellulose takes its name [25] from the French chemist Anselme Payen (1795–1871) who suggested in 1831 that the cell walls of all plants have inside a common substance, which he just called cellulose. The easiest way to describe the chemical structure of cellulose is probably to imagine to build it "piece by piece" (Fig. A.1).

The only basic molecule required to assemble cellulose is glucose, a sugar having a hexagonal ring structure with five carbon atoms and one oxygen atom. If we join together, with an oxygen atom as a bridge, two molecules of glucose, but having before rotated one of the two glucoses upside down with respect to the plane of the hexagonal ring, we get the "fundamental brick" that constitutes the cellulose.[1] Cellulose turns out by the union with oxygen bridges of many of these identical basic building blocks, even up to 10,000 [131]. Thus cellulose is a polymer[2], more specifically a

---

[1]It is the molecule called cellobiose; from the chemical point of view it is a "disaccharide," because it is formed by the union of two ("di") sugars ("saccharide").
[2]See footnote 6 on p. 168.

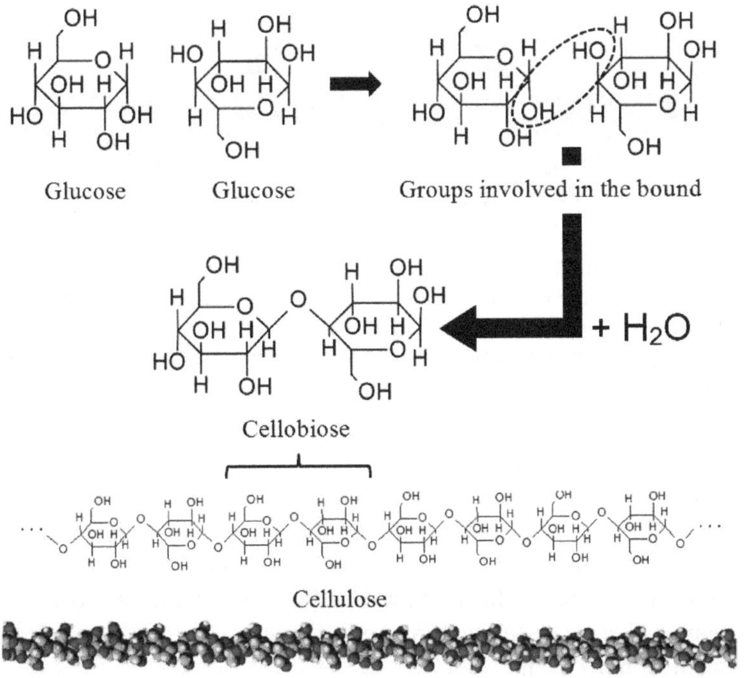

Glucose   Glucose   Groups involved in the bound

$+ H_2O$

Cellobiose

Cellulose

Figure A.1 Schematic draw describing the chemical structure of cellulose. The cellulose's "fundamental brick" (the cellobiose) is obtained by the union of two glucose molecules (one of which previously put upside down with respect to the plane of the hexagonal ring) by means of an oxygen atom that plays the rule of a bridge. The cellulose is then obtained by the union of a very large number, even up to 10,000, of these basic building blocks; this gives to cellulose the structure of a long chain without lateral ramifications.

polysaccharide,[3] composed only of the elements carbon, hydrogen, and oxygen. But its characteristics do not end here.

The presence of OH groups in the glucose molecule makes possible the formation of particular weak bonds, the hydrogen bonds,[4] between the cellulose chains, which constitute hooking points. Thanks to these special interconnections the cellulose chains are also able to set parallel to each other for long sections in a

---

[3]See footnote 7 on p. 168.
[4]Also known as "hydrogen bridge," it is the most energetic of the so-called weak bonds.

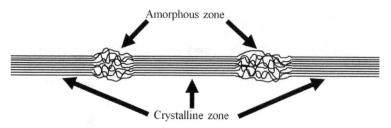

Figure A.2 Simplified drawing representing the crystalline and amorphous disposition of the cellulose chains. In the crystalline regions the polymer chains are arranged in a compact and orderly way with precise dimensions (length and width) with which each chain is separated from the others all around.

very compact way (Fig. A.2). This generates a three-dimensional tidy and ordered configuration in which the hydrogen bonds are much more numerous than those present in a random arrangement of the cellulose chains. In the first case (ordered spatial disposition), cellulose has a crystalline form, in the second case it is said to have an amorphous form.

In vegetable fibers cellulose is present in both forms and this greatly affects the mechanical properties of the different structures containing it into the plant. We can easily imagine that a stretched structure formed in large part by crystalline cellulose (namely with chains all parallel and close to each other) deforms itself less than a structure having the same number of chains but arranged in a more disordered manner (amorphous cellulose). In fact in the second case the chains can more easily slide each other under tensile forces, because there are less constraint points realized with hydrogen bonds.

The so-called *microfibrils* play an important task inside a vegetable fiber. They are thread-like agglomerates formed by 30–100 cellulose molecules arranged in a crystalline form [87] and characterized by diameters ranging from 2 to 20 millionths of a millimeter (from 79 to 790 billionth of an inch) and lengths varying from 0.1 to 40 thousandths of a millimeter (from 3.9 to 157 millionth of an inch). [6] The mechanical strength of a single flax fiber (in other words the ability to withstand traction) depends on the microfibrils.

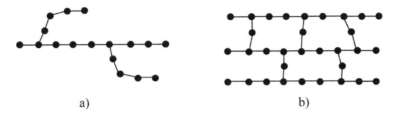

a)   b)

Figure A.3 (a) Representation of a branched polymer: lateral branches rise from the main chain. (b) Schematic draw of a cross-linked polymer: its chains are joined by other (like bridges) and this gives rise to a three-dimensional net structure. The interconnection chains have a variable number of basic building blocks.

A.5 Cap. 5: The Other Fundamental Substances in Plant Fibers

After cellulose (Table 5.1) the basic constituent substances present in vegetable fibers are in order hemicellulose, lignin, and pectin. We report their salient features below. Although there are other minor substances in the vegetable fibers, including waxes and ashes, they are not described here.

Hemicellulose

Hemicellulose is not a different form of cellulose (Section A.4), but it is a polysaccharide[5] consisting of different basic building blocks (namely sugars) joined together in number [25] 10 or even 100 times smaller than cellulose. Since [6] different sugars are bound in the hemicellulose polymer, this means that it is not strictly the same from plant to plant. In addition, [134] the hemicellulose chains have ramifications and have distinctly amorphous form; so hemicellulose is a branched polymer (Fig. A.3a) and hydrophilic, while the cellulose molecule is hydrophobic.

Like the plastic sheath that wraps the electric copper wires, hemicellulose establishes hydrogen bonds [94] with the superficial cellulose chains of the microfibrils; thus, it connects the adjacent microfibrils by covering their surface.

---

[5] See footnote 7 on p. 168.

Lignin

The name of this compound is due to the Swiss botanist Augustin Pyramus de Candolle (1778–1841), who proposed it in 1819 from the Latin word *lignum* (wood). After cellulose, lignin is one of the most abundant natural organic substances on Earth whose chemical characterization is particularly complex and [9] with some obscure points.

Lignin [94] is a highly branched polymer with only carbon, hydrogen, and oxygen and with the chains interconnected through strong bonds (Fig. A.3b); this gives rise to a three-dimensional lattice, a hydrophobic structure, highly resistant [87] to microorganism attacks, which helps [134] to give structural rigidity to the plant, especially during the maturation phase, where the stiffening becomes maximum.

Pectin

Pectin is actually a family of polysaccharides [94] with branched chains. Pectins are hydrophilic like hemicellulose and therefore capable of absorbing water.

A.6 Cap. 5: The Internal Structure of a Flax Fiber

In Fig. A.4 we report a schematic drawing of the concentric multilayered structures inside a flax fiber, which we would gradually meet imagining to peel it with a knife.

The structure called *middle lamella*, whose main constituent material [99] is pectin followed by hemicellulose[6] (Section A.5), works as a glue between adjacent fibers in the bundles inside the phloem (liber) of the flax plant (Figures 5.1 and 5.2).

In a particular phase (retting, Section 5.2) of the extraction process of the flax fibers from the stem the disintegration of a gradually increasing number of pectin's polymeric chains into the middle lamella takes place, so finally the fibers turn out to be free. Since the middle lamella is broken up, in the literature it is not

---

[6]These substances are both hydrophilic.

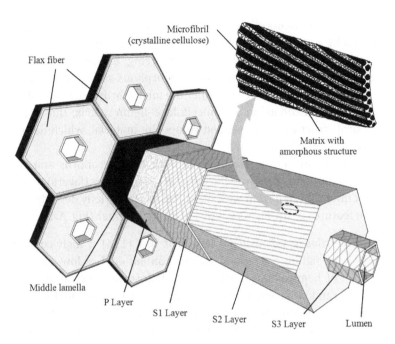

Figure A.4 Schematic representation of the concentric multilayered structures into a flax fiber assuming a hexagonal regular cross-sectional model; the names of the various layers follow those in literature. The organization of cellulose microfibrils (which are highly crystalline structures) within the thickest layer S2 has a particular importance for the mechanical properties of a flax fiber. The microfibrils are <<wound in a helix>> and immersed in a matrix with amorphous structure composed of hemicellulose and lignin. The schematic drawing in the upper right represents this particular structure in a simplified way.

normally considered part of the structures constituting a single fiber flax. In Fig. A.4 we also represented the middle lamella.

Moving from the outside toward the axis of the fiber, we meet in order the following coaxial structures: the primary cell wall (P layer), the secondary cell wall (S layer), and a hollow duct, the *lumen*, which is an open channel at the center of the fiber and which passes it from one side to the other, facilitating [131] the transport of nutrients. In the flax fiber the lumen generally occupies less than 10% of the cross section [35] and may be [139] as small as 1.5%.

All types of lignocellulosic cell walls, linen included, are characterized by a common architecture [142]: the cellulose microfibrils[7] (Section A.4) play the role of reinforcing elements within a polymeric hemicellulose and lignin amorphous structure. In other words, in the cell walls of a flax fiber the microfibrils have crystalline cellulose "embedded" in a hemicellulose–lignin matrix; this is the classical structure of a so-called composite material, which nature "invented" long before man.

The amount and distribution of both the microfibrils and the constituents of the matrix make the coaxial structures (layers) inside a flax fiber different from each other. Let us now briefly describe the main features of the layers P and S inside a flax fiber (Fig. A.4).

- *Primary cell wall* (P layer): it has [114] a high content of lignin, a medium pectin content and a low amount of hemicellulose and [139] cellulose with predominantly random orientation. With a thickness [18] of about 0.0002 mm (0.0000079 in.), it is less than 2% [131] of the total thickness of the fiber. As during retting (Section 5.2) pectins are attacked by enzymes, an excessive degree of retting may involve the P layer, damaging its structures.

- *Secondary cell wall* (S layer): it represents the 90% [131] of the thickness of a flax fiber and is commonly believed to be divided into three sublayers. Proceeding from the outside to the fiber axis, such sublayers are indicated in the literature with the abbreviations S1, S2, and S3.[8]

  The thickest of the three substrates is the S2 one, which can reach up to 80% [139] of the fiber cross section; consequently this sublayer is the one that most influences the mechanical properties of the flax fiber, in particular its stiffness and tensile strength, that is, when the fiber is stretched parallel to its axis. The S2 sublayer has a

---

[7] They have a crystalline form.

[8] The letter S is the shorthand of the word "secondary," because this structure is set into a more external one, called primary (the layer P in Fig. A.4). The abbreviations S1, S2, and S3 were introduced for the first time by I. W. Bailey in 1935. We observe that in the literature, still going toward the axis of the fiber, the order of the layers can often be written in the opposite way, that is, S3, S2, and S1—hence the importance of referring to a schematic drawing in order to dispel any doubts of interpretation.

high content of cellulose with a crystallinity level that can reach 90% [131] (which is related to the high presence of microfibrils), a medium content of hemicellulose, and a low presence of lignin [114].

The microfibrils in the matrix are spirally wound as a levorotatory[9] helix characterized by a winding angle of around [18] 10°with respect to the axis (Fig. A.4).

As regard sublayers S1 and S3, they have [114] a smaller amount of cellulose compared to the S2 one and a greater presence of the other fundamental constituents that form the matrix. Particularly interesting is that in them the microfibrils are not only levorotatory helically wound as in sublayer S2, but both levorotatory and dextrorotatory with a winding angle of around 70°; this gives rise to a kind of network that avoids any twisting and limits the variations of thickness of the layer during the stretching of the fiber [80].

It is worth noting that the same coaxial structures are present in many cellulosic plant fibers, but with different thicknesses of the layers (even if sublayer S2 remains the thicker and to which the stiffness is connected) and winding angle of the microfibrils.

A.7 Cap. 5: The 3–1 Shroud's Weaving Technique

The Shroud has a unique design with a herringbone band width of about 11 mm (0.43 in.) because it has been diagonally weaved with the three-over-one technique. Fig. A.5 schematically shows the three-over-one technique in reference to a weaving procedure carried out on a vertical handloom of 2000 years ago.

---

[9]The term describes the winding direction of the helix: to ideally trace a levorotatory helix in space with the tip of a finger we have to turn counterclockwise when going up. If, always going up, the tip of the finger rotates clockwise, the helix is said dextrorotatory instead. In the case of twisting of the textile yarns and by extension also to describe the winding of the microfibrils in the sublayers of the plant fibers, the terms "Z" twist (levorotatory helix) or "S" twist (dextrorotatory helix) are also used (Fig. 5.6).

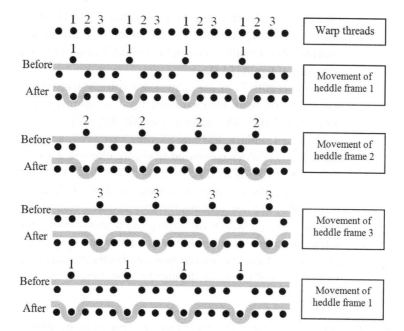

Figure A.5 Schematic drawing of the three-over-one weaving technique. All the warp yarns with the number 1 above are moved by the heddle frame number 1, those with the number 2 above by heddle frame 2, etc.

The warp threads are represented in section (black circles). Some warp yarns are always fixed; in other words they are not moved by the heddle frames (circles without number above). Other warp yarns are moved from their rest position by means of heddle frames and each heddle frame moves a set of yarns. To achieve the weave of the Shroud the vertical handloom must have had three heddle frames (realized with horizontal rods).

In the manual weaving of 2000 years ago this sequence was cyclically repeated.

(I) With the heddle frame number 1 the corresponding warp threads were raised, the weft yarn passed through, and then the warp closed. Every four warp threads, the weft thread turn out to be alternately set above three and below one warn yarns.

(II) With the heddle frame number 2 the corresponding warp threads were raised, the weft yarn passed through, and then the warp closed. Like before every four warp threads the weft thread is alternately set above three and below one warn yarns but compared to the previous step a shift to the right of one warp yarn is also done.

(III) With the heddle frame number 3 the corresponding warp threads were raised and the weft yarn passed through in the same way and with the same shift to the right.

A.8 Cap. 5: The Cross-Polarized Light Technique

Here we shortly explain in a quite technical way the reason why by observing the flax fibers by means of a petrographic microscope they show a special coloring.

First of all we remember that in physics the polarization of the electromagnetic radiation (and therefore also for the light rays) is a characteristic of the electromagnetic waves: a light beam can be thought like a set of photons that travel in vacuum at the speed of 299,792,458 m/s (186,282 miles per second) and oscillate along a plane otherwise inclined. Their various wavelength produces a different color of the light source.

If we interpose a polarizing filter perpendicularly to the light beam, it produces a selection (filtering) of all the incident photons. The filtering consists in letting pass through the filter only the photons that oscillate along a particular plane, the so-called plane of polarization.

The light photons normally oscillate in a completely random way; thus there is not a preferred direction of oscillation. In the drawing in Fig. A.6 we depicted only two particular oscillation directions of the photons, perpendicular to each other, among all those possible. Let us now imagine that the polarizing filter, on which the light strikes perpendicular to the plane, has vertical polarization. To the photons of the light beam the following occurs:

- Those that oscillate vertically pass undisturbed through the filter.

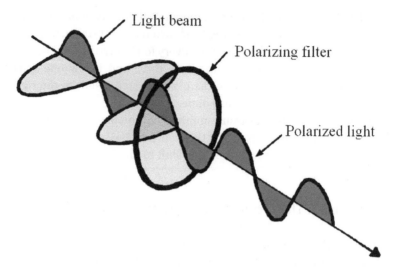

Figure A.6 Operation of a polarizing filter with the polarization plane supposed vertical. A ray of light strikes perpendicularly on the filter: for simplicity, among all those present in light beam, only two particular oscillation directions of the photons, perpendicular to each other, have been depicted. The filter is passed unchanged by photons that oscillate vertically; it totally blocks those horizontally oriented and reduces in intensity all the other photons, by orienting them vertically along the plane of polarization.

- Those that oscillate horizontally are not able to pass the filter, because their direction of oscillation is perpendicular to its polarization plane.
- Those with oscillation directions intermediate between the two preceding pass the filter attenuated in light intensity and leave it with a vertical plane of oscillation.

The filtering has a "cost": the light beam emitted from the filter has a lower light intensity because many photons are absorbed by the filter, namely they cannot cross it. However, what matters at the application level is that the photons coming out from the filter all oscillate along a plane having a known inclination, even if at the price of a reduction in intensity of the outgoing light beam. Because of the particular oscillation that characterizes the photons of the electromagnetic waves coming out from a polarizing filter, they are called *polarized* waves. At this point, what happens if the

Figure A.7 Effect of the rotation of the polarization plane of a filter, which is perpendicularly stricken by polarized light produced by a source not visible in the photographs. The angle between the planes of polarization increases from left to right, with the result that step by step a smaller number of photons are able to pass the filter (the light intensity decreases). When the polarization planes are perpendicular to each other, the incident light does not pass through the filter (the so-called night effect, which is exploited by photographers).

polarized light emerging from the filter strikes perpendicular to another polarizing filter?

The result depends on the mutual orientation of the polarization planes of the filters (Fig. A.7). We said that the first filter directs (filtering) the photons of a light beam, making them oscillate along a particular plane; the second filter further filters the same photons along a different plane. If the plane of polarization is perpendicular to that of the first filter, theoretically all the photons of the light beam are absorbed. Even photographers exploit polarizing filters to reduce glare or to apply the so-called night effect, which is obtained just by rotating the filters in such a way that their planes of polarization are perpendicular to each other. So, by adjusting the mutual inclination of the polarization planes of the two filters, we can control the intensity of the outgoing light beam, since the photons can more or less cross the second filter. It is on these optics principles that a petrographic microscope works so as to be able to highlight a particular characteristic of the flax fibers.

Let us suppose that we have a light beam polarized along a vertical plane. The beam is composed of several photons, each oscillating with a certain wavelength (color) but out of phase[10] with

---

[10]The phase is a physical characteristic associated with waves, light included. We try to explain this concept by an analogy. Imagine two perfectly identical pendulums put into oscillation. The pendulums make a round and a return (oscillation) exactly identical, but compared to the first, the second pendulum can reach the position

respect to one another. It turned out that the different characteristics of the flax fiber change in phase the path of the incident photons proportionally to the magnitude of the characteristics themselves. This occurs because the flax fibers have a particular optic characteristic (birefringence) capable of splitting a striking light beam into two differently polarized rays.

To better explain this concept just imagine a flax fiber as a large cylinder of Plasticine. Beating its surface with a stick with sufficient intensity, the signs of the caused damage will be etched on it like transverse depressions: the material has permanently been deformed. The greater the permanent deformation (damage) stored in the fiber, the greater the phase shift that the material gives to the incident photons with respect to those hitting over neighboring areas gradually less deformed or remained intact. So in the phase shift of the outgoing photons information of the fiber's characteristic (the plastic deformation in this example) is contained.

But the human eye is not capable of perceiving the phase shift of the various photons and so it is not able to observe particular properties related to this phenomenon. It is therefore interesting to find a method that allows the human eye to "see" these phase shifts and in this way to characterize further physical properties of a flax fiber. It is at this point that the petrographic microscope turns out to be very useful.

A petrographic microscope has two polarizing filters with which we can detect the phase shifts of the photons and so study the flax fibers. In fact, if we use a polarizing filter under the slide containing the samples and a second polarizing filter (called analyzer), placed over the slide and rotated 90°compared to the previous one, we obtain the darkening of the whole visual field (due to the night effect described above), while some details of the fibers shine with a particular coloring also dependent on their physical structure. It is in

---

of maximum deviation from the vertical before (in advance of phase), at the same time (in phase), or after (late in phase). Thus the phase quantifies the advance or the delay between two oscillations: the phase is zero only in the case where the two pendulums reach the position of maximum deviation from the vertical at the same instant.

this manner that the changes in physical properties, resulting from the damage induced on the surface of the fibers, can be studied.[11]

At the end, here it is an easy way to distinguish in the Shroud's so-called vacuumed dusts the ancient fibers from the relatively modern ones—by seeing them under a petrographic microscope to observe and studying the extent of defects. Between the various flax and other kind of fibers, there are those belonging to the Shroud and others relatively modern (such as those coming from the sixteenth-century Holland cloth); the Shroud's fibers are easy to recognize, thanks to the special coloring that they assume compared to fibers that have much less than a millennium.

A.9 Cap. 7: Elasticity and Plasticity

Elasticity and plasticity or, better, elastic and plastic behaviors are two fundamental concepts of materials. For simplicity, here we consider a thread, even if these concepts apply to bodies of any shape.

By elasticity is meant the property of a material to stretch when pulled, so when the stress ends, the body always returns with exactly the same dimensions that it previously held. This behavior is explained at the atomic level in Fig. A.8.

On the other hand, we use the term "plasticity" referring to the property of a material to stretch without regaining its original size after the stress has been removed. In other words, a residual and permanent deformation (so-called plastic for this reason) remains into the material. A clay cylinder exactly behaves in this manner, if pulled. Figure A.9 explains the plastic behavior at the atomic level.

We can find elastic and plastic behaviors even in complex materials, such as fibers, depending on the applied stresses. In the case of a flax fiber, to mutually slide or not slide are the polymeric chains, first of all those of cellulose (Section A.5), inside the multilayered structures described in Section A.6. From the microscopic point of view some hydrogen bridges[12] break away and

---

[11] By the way this could be another dating method to be developed in the near future.
[12] See footnote 4 on p. 375.

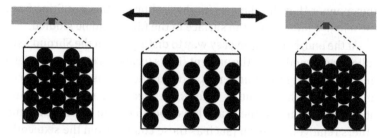

Figure A.8 Schematic drawing to explain the elastic behavior at the atomic level. From a situation of rest (on the left), the body is pulled (middle): the atoms move away from each other with respect to their equilibrium positions, but never breaking away or sliding each other. At the macroscopic level the body stretches. Once the stress has been removed (on the right), the atoms return to their original positions and the elongation completely disappears.

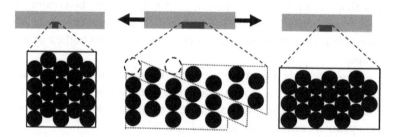

Figure A.9 Schematic drawing to explain the plastic behavior at the atomic level. From a situation of rest (on the left), the body is pulled: the atoms move away from each other with respect to their equilibrium positions, but at the same time some sets of atoms slide mutually, changing their position (center). At the macroscopic level the body stretches. Once the stress has been removed (on the right), the atoms recover the elastic elongation, but those that have widely sliding moved are not able to return to their original position. This explains the presence of a residual and permanent elongation in the body.

form again between different portions of the polymeric chain, which in the meantime has been able to slip with respect to the other nearby chains. This explains the presence of an eventual residual and permanent macroscopic elongation shown by a stretched flax fiber.

A.10 Cap. 7: Force Measuring Feasibility with an Analytical Balance

In Fig. A.10a we report a schematic drawing of the experimental apparatus that has been made by the second author to verify the possibility of employing a commercial analytical balance[13] to measure the tensile forces acting on a single stretched flax fiber.

The micrometer slide (Fig. 7.4b), namely the displacement reducer, has been fixed by means of clamps to two stanchions, disposed one at each side[14] (they are not shown in Fig. A.10a). We chose a common weight of 100 g (3.53 oz.) to connect the tabbed flax fiber to the balance plate, while at the first moment we used as a tab (Fig. A.11a) a rectangular portion of common drawing paper with a rectangular notch of about 1×7 mm (0.039×0.27 in.). The top and bottom parts of the tab were connected by two lateral bridges of roughly 1.5 mm (0.059 in.) in width each.

To extract single flax fibers from the yarns and to tab them by means of a common cyanoacrylate glue we used a stereo-optical microscope (Fig. A.10b) with variable magnification from 7X to 70X.[15]

The tab with the single flax fiber was glued on one side on the lateral surface of the weight and on the other side in correspondence to the bolt's head. Then, in the absence of traction, we cut the lateral bridges with two scissors, thus leaving only the fiber connecting the two tab parts.

We unfortunately found this first tab model completely unusable, because while cutting the lateral bridges with scissors the stresses transmitted to the fiber due to the cardboard's thickness were of such entity to instantly break the sample. Consequently we had to change the tab type. We then used a portion of the side strips

---

[13]The analytical balance was kindly loaned to the second author—with other materials—by the Marco Foscarini High School (Venice). This balance has a maximum load capacity of 150 g (5.29 oz.) with a resolution of 0.005 g (0.00018 oz.).

[14]Due to the presence of the tripod base of the stanchions, it has not been possible to put the balance plate directly under the micrometer slide, so a pin, made with a bolt, has necessarily been used.

[15]The most used magnifications were 20X and 40X.

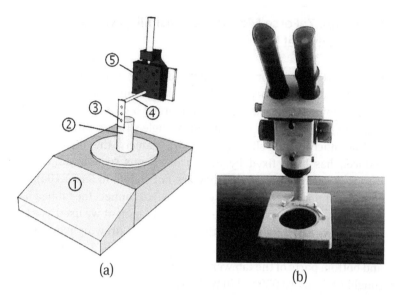

(a)   (b)

Figure A.10 (a) Schematic drawing of the experimental apparatus used to perform a preliminary test for measuring the force acting on individual flax fibers from both modern and antique fabrics. (1) Analytic balance, (2) weight, (3) tab with the fiber (in correspondence to the central hole), (4) pin (bolt), and (5) micrometer slide. (b) Stereo-optical microscope (with variable magnification 7X–70X) used to extract the single flax fibers from the yarns and to tab them.

of a continuous-form printing paper[16] (Fig. A.11b). In this case the cutting operation of the lateral bridges turned out to be less traumatic for the fiber, although still not easy. However, because of the size of the holes, it would have been impossible to use the tape model tab on single ancient fibers 1–2 mm in length. For this reason we had to consider a different tab type to enable us to test on fibers ranging within such short lengths.

Taking inspiration from the tab that had been utilized by the Max Planck Institute in 2009 (Fig. A.12a), we decided to use a transparent polymeric and readily available material such as that of a common A4 polyester sheet for photocopies.

---

[16] Each tab (14 × 65 mm; 0.55 × 2.56 in.) coming from such a strip had five holes. Since the diameter of each hole was 4 mm (0.16 in.), therefore only single flax fibers at least of 5–5.5 mm (0.20–0.22 in.) in length could be tabbed.

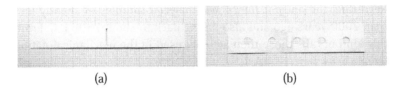

(a)   (b)

Figure A.11 (a) The first model of a tab (size 10 × 70 mm; 0.39 × 2.7 in.), which has been processed from a drawing paper sheet; (b) "tape" model tab (size 14 × 65 mm; 0.55 × 2.56 in.) derived from the side strips of a continuous-form printing paper. The flax fiber (invisible in the picture) has been glued in correspondence to the middle hole.

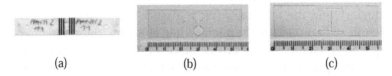

(a)   (b)   (c)

Figure A.12 (a) The tab used by the Max Planck Institute to perform the tensile tests on the samples sent in 2009; (b) the first polyester tab coming from a common A4 photocopy sheet; (c) a later version of the tab shown in (b) with an H-shaped carving, from which the tabs used in our tensile tests actually came.

The first model of a polyester tab was realized, as shown in Fig. A.12b. The circular holes were made with a paper sheet–drilling machine, thinking to speed up the processing times. But after a few tests, we observed that the insufficient sharpening of the machine's punches and the plasticity of the support gave very poor quality tabs. We then realized an H-shaped handmade carving with a cutter (Fig. A.12c). This different solution gave, even if with an increased production time, tabs of better quality and with a greater length of the horizontal portion on which to glue the ends of the flax fiber.

To break up the lateral bridges we thought to fuse them by using a red-hot wire so as to transmit to the tabbed fiber the least stress amount, especially of the shear kind, being these the main cause of the fiber's failure at this step of the experimental procedure.

With such polyester tabs, fixed to the experimental device shown in Fig. A.10 in the same manner used for the tape model tabs, we performed some tensile tests on modern and ancient fibers.

Regarding the ancient fibers, we analyzed samples of slightly less than 2000 years ago and also dating back to about 2400 B.C. From this kind of tabs those then actually used in the mechanical dating tensile tests were realized.

A.11 Cap. 7: Metrological Problems and Calibration

The calibration involved two elements of the microcycling tensile machine, the displacement reducer and the analytical balance.

As for the displacement reducer, the real cantilever beam is constrained through a system that realizes an imperfect fixed support, and the machine's frame elastically deforms during the test in addition. To use the microcycling tensile machine what must be known (static calibration) is the real relationship between the displacement of the cantilever beam's free head and the displacement of the hook on which the tab with the sample is hanged. To achieve this goal we recorded the hooks' displacements (Fig. A.13a) by means of a comparator with the following characteristics: range 0–5 mm (0–0.20 in.) and resolution 0.001 mm (0.000039 in.).

For the hook with a nominal reduction ratio equal to 1/20, the only one used to perform the tensile tests on flax fibers, the experimental (in other words, the real) reduction ratio turned out to be 1/18.9 ($0.0528 \pm 1.7\%$).

As regard the analytical balance, its load cell is a monobloc magnetic compensation type. The construction details of this balance's device are a trade secret, but the operating principle and the basic functional components of a monobloc magnetic compensation load cell weighting system can be gathered from patents. For the analytical balance we considered the following aspects:

(i) Its total stiffness
(ii) Its behavior over time

The balance's total stiffness is a measure, even if invisible, of how its plate is lowering when the instrument is weighing. If the plate moves, this means that the extension of the flax fiber is composed of two terms, one related to the change in length of the fiber when

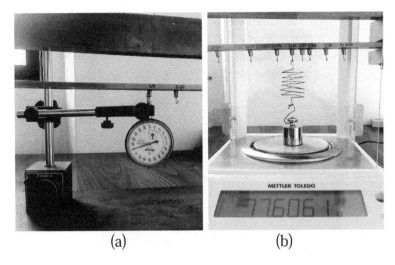

Figure A.13 (a) Example of measuring the hooks' displacements by mean of a comparator (the picture refers to the acquisition of the values of the section with nominal reduction of 1/5); (b) analysis of the behavior over time of the analytical balance when the mass to be weighted is connected to a frame through a spring (which simulates the flax fiber). With the same experimental setup we have studied the correctness of the results provided by the balance used in such a different way than that provided by the manufacturer.

stretched and the other due to the plate displacement. Since a force corresponds to an extension, it follows that the measured force is affected by an error connected with the insertion of the measuring instrument in the system (this is known by the technical term "loading effect"). If, however, the loading effect is negligible, as experimentally verified, then the readings given by the balance are the actual force acting on the fiber without the need for any subsequent correction.

As during the tensile test the tabbed fiber is constrained (Fig. 7.10) on the one side on one of the hooks of the displacement reducer and on the other side to the additional mass resting on the analytical balance's plate, it turns out that the weighing instrument is not used, as provided by the manufacturer.

In fact a stretched fiber behaves like a spring connecting the additional mass on the plate to the frame of the microcycling tensile

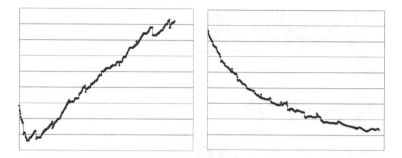

Figure A.14 Diagrams for two different load conditions in which an apparent mass variation (drift) is evident when the weight on the plate is as shown in Fig. A.13b. The displayed values have been recorded every 3 seconds for a total time of 1000 seconds; on the ordinate axis the values range from 77.753 g (2.7427 oz.) to 77.762 g (2.7430 oz.).

machine. So the mass on the plate is therefore not free, as provided in the usual working way of the analytical balance.

We built a handmade steel wire spring to simulate a flax fiber. Then we studied the behavior of the analytical balance over time by placing on its plate a weight of 100 g (3.53 oz.) and then one of 50 g (1.76 oz.) connected to the displacement reducer (hook with a nominal reduction factor of 1/20) through the spring (Fig. A.13b). In such a way we obtained the experimental proof that drift is present when the mass to be weighed is not free (Fig. A.14). The behavior shown by the analytical balance used in this particular manner is due to the unprovided spring (namely the flax fiber during the tensile test) that interferes with the force compensation magnetic system of the load cell of the instrument.

What we just pointed out made it necessary to verify the correct working of the analytical balance by answering the following three questions:

(1) Are the data provided by the analytical balance correct in terms of values read on its display?
(2) Does the response of the instrument still remain linear?
(3) Is the response still linear under different stiffness values of the springs (namely under different flax fibers) connecting the balance's plate to the microcycling machine's frame?

Table A.2 Operating characteristics of the microcycling machine relative to the hook with a nominal reduction factor of 1/20, the only one used to perform the tensile tests on flax samples

Force	Displacements	
Uncertainty (95%)	**Range**	**Uncertainty (95%)**
±0.001 cN (±0.0000022 lbf)	0.000–0.050 mm (0.000–0.0020 in.)	±0.0050 mm (0.00020 in.)
	0.050–0.300 mm (0.0020–0.012 in.)	±0.0060 mm (0.00024 in.)
Operating range	0.300–0.400 mm (0.012–0.016 in.)	±0.0070 mm (0.00027 in.)
0–50 cN (0–0.11 lbf)	0.400–0.500 mm (0.016–0.020 in.)	±0.0085 mm (0.00033 in.)
	0.500–0.600 mm (0.020–0.024 in.)	±0.0100 mm (0.00039 in.)
	0.600–0.700 mm (0.024–0.027 in.)	±0.0110 mm (0.00043 in.)
	0.700–0.792 mm (0.027–0.031 in.)	±0.0125 mm (0.00049 in.)

Our experimental results provided an affirmative answer to all these three questions. So, even if the flax fiber interferes with the analytical balance's measurement system calibrated in factory, notwithstanding the drift, the values provided by the instrument are nevertheless reliable within the limits of the experimental errors provided that each value is recorded within two to three seconds. In fact in such a way the drift's error affecting each measured value does not exceed ± 0.0005 cN (± 0.0000011 lbf).

Thanks to the calibration of the microcycling tensile machine we can state that the instrument is capable of measuring forces with a resolution of 0.0002 cN (± 0.00000045 lbf), as provided by the design specifications, and with a maximum load capacity of 50 cN (0.11 lbf). It is furthermore possible to increase the maximum load capacity up to 70 cN (0.16 lbf).

For the hook with a nominal reduction factor of 1/20 the calibration values are summarized in Table A.2.

A.12 Cap. 7: Experimental Tensile Tests

We followed an experimental procedure divided into the following steps:

(1) Selection and tabbing of a single flax fiber (sample)
(2) Execution of pictures of the tabbed sample

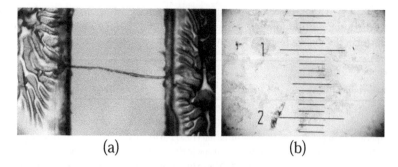

(a)   (b)

Figure A.15 (a) Example of a flax fiber's photograph at 40X in polarized light to better highlight it (sample A.07); (b) image (electronic caliper) at 40X of the millimeter grid in vertical view (an horizontal one is also performed) that accompanies each lot of pictures of tabbed samples.

(3) Gluing of the additional mass and transferring of the sample on the tensile machine

(4) Execution of the tensile test

We described in Section 7.3 the procedures (both carried out with the aid of a stereomicroscope) of the flax fibers' preliminary selection and of their extraction from a yarn. Let us here only recall that the single flax fiber is not-taut tabbed with cyanoacrylate glue on our handmade mask in order to prevent rupture during the handling of the sample and in the delicate operation of melting the two lateral bridges of the tab.

We performed pictures of the single tabbed flax fibers (Fig. A.15a) for all the prepared lots of samples. Each fiber is photographed in polarized light at a magnification of 20X or 40X. We decided to take at the same time a vertical and a horizontal digital view of a millimeter grid (Fig. A.15b) with the same magnification factor used for photographing each lot of samples. This was done to obtain an electronic caliper, which we used to measure the gauge length of the fiber and its diameter from the image, because these are the dimensions requested to obtain the stress–strain curve.

The tab with the fixed sample on it has to be glued within two steel blocks (Fig. A.16a) to provide it with an additional mass, which constrains the polyester mask's lower part to the analytical balance's plate during the tensile test. Then the whole system is rotated in

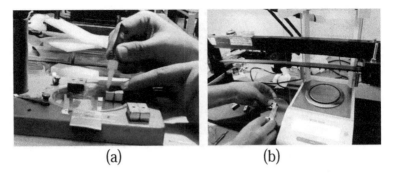

(a)   (b)

Figure A.16 (a) A frame depicting a step of the gluing of the additional mass onto the tab. The lower block is already fixed and the second author is depositing a cyanoacrylate glue drop on the polyester mask, stopping the lower block with a finger. (b) Rotation manner with which the tab with the additional mass (indicated by the arrow) is moved from the horizontal to the vertical position with the minimum stress on the fiber and simultaneously moving the whole system inside the weighing chamber of the analytical balance.

a vertical position and brought with caution inside the weighing chamber (Fig. A.16b), avoiding as much as possible to stress the fiber, especially if it is very ancient.

To perform the tensile test, first of all the lateral bridges of the tab must be cut away (Fig. A.17). This operation turned out to be quite delicate and it required the acquisition of a certain manual dexterity. In fact several samples were broken at this stage (Table 7.3) because of the stresses inadvertently transmitted to the tabbed fiber.

Each tensile test was performed in this manner.

Since moisture affects the mechanical properties of the flax fibers, before starting the test we provided to record on a special data sheet the relative humidity measured with a digital hygrometer with a resolution of a percentage unit. Then for each tested sample we collected the various couples of *imposed displacement/weighed mass* values.

Once the tensile test was concluded, we rejoined the two sides of the tab with Scotch tape (taking care not to damage each part of the broken fiber) and then we placed the specimen into a plastic envelope.

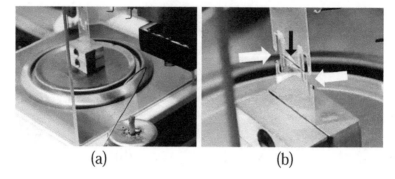

(a)   (b)

Figure A.17 (a) A frame of the cutting operation of the tab's lateral bridges. The end of a steel wire suitably bent is red-hot-heated, holding it with a clamp, on the upper part of the flame of a handmade alcohol lamp. After that, this part of the wire is gently moved closer without contact to the lateral bridge to be cut by fusion. The cutting is made in a few steps, never all at once. (b) Detail of a tab with the lateral bridges definitely cut away (white arrows). At this point only the fiber (below the black arrow) is connecting the two halves of the tab and then the tensile test can begin.

A.13 Cap. 7: Definitions of the 7 Measured Mechanical Parameters

We evaluated seven mechanical parameters from the $\sigma-\epsilon$ standardized diagrams of each successfully tested sample, which had furthermore passed the two exclusion criteria (d) and (e) (Section 7.7) that we decided to add with the aim of minimizing the data scattering.

Below we give the definitions of these mechanical parameters.

(1) Breaking Strength, σ_R

The breaking strength, σ_R, is the ordinate of the absolute maximum of the $\sigma-\epsilon$ curve (Fig. A.18a).

(2) Elastoplastic Strain at Break, ϵ_R

The elastoplastic strain at break, ϵ_R, is calculated as the difference between the value of the abscissa of the breaking point and the abscissa corresponding to $1/100$ of the breaking strength σ_R.

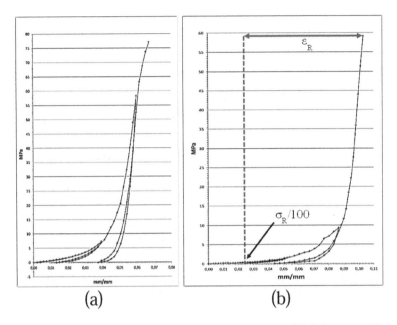

Figure A.18 (a) Definition of the breaking strength, σ_R; (b) definition of the elastoplastic strain at break, ϵ_R.

Figure A.18b graphically shows the definition of this mechanical parameter, which quantifies the elastoplastic deformation of the fiber.

(3) Final Elastic Modulus, E_f

The final elastic modulus, E_f, is a Young's modulus derived from that specified in [14] but that takes into account the presence of the cycles (Fig. A.19).

(4) Elastic Strain at Break, ϵ_E

With this mechanical parameter, connected to the final Young's modulus, E_f, we quantify the elastic deformation of the samples; it is in fact computed in the zone with a predominantly elastic behavior of the fiber.

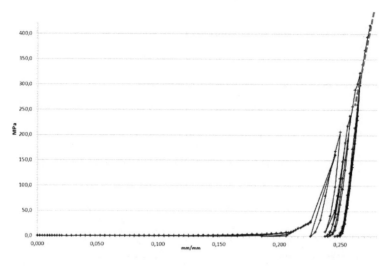

Figure A.19 Indication of the portion of the σ-ϵ standardized diagram in which the final elastic modulus, E_f, is calculated as the angular coefficient of the straight line interpolating the points.

The elastic strain at break, ϵ_E, is the difference between the abscissa of the absolute maximum point of the σ-ϵ curve and the intersection point of the straight line interpolating the points relative to the Young's final modulus, E_f. Figure A.20a shows its practical computation.

(5) Decreasing Elastic Modulus, E_c

A cycle is defined as any sequence of a loading phase followed by an unloading phase (Fig. A.21a). For a flax fiber the slope of the loading curve of a cycle is distinct from that of the corresponding unloading curve. This clearly indicates the presence of hysteresis (internal energy dissipation).

The tested flax fibers show a stiffness increase with the number of cycles that reflects what has already been experimentally observed by C. Bailey [18]. This behavior is supposed to be connected to the changes of the microfibrils angle (Section A.6) [100] that produces a packing effect [53] on the fiber. So it seemed appropriate to take into consideration the different slopes of the two

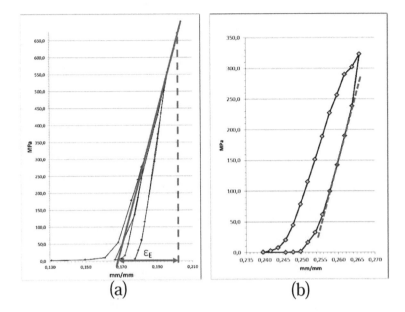

Figure A.20 (a) Evaluation of the elastic strain at break, ϵ_E, from the $\sigma-\epsilon$ standardized curve; (b) computation method of the decreasing elastic modulus, E_c, as the angular coefficient of the straight line interpolating the linear portion of the descending part of the direct cycle with the maximum stress.

curves that form a single cycle through the definition of a different elastic modulus indicated as E_c. The decreasing elastic modulus, E_c, is a particular Young's modulus that we introduced to quantify the stiffness of the fibers in the presence of cyclic loading conditions. This mechanical parameter is defined considering the unloading part of the cycle with the maximum stress (Fig. A.20b). In this cycle's section the plastic component (Section A.9) turns out to be on average smaller, so the elastic modulus, E_c, is less subject to the effects of the fiber's plasticity.

(6) Direct Loss Factor, η_D

Under cyclic loading conditions a flax fiber shows an internal energy dissipation, the so-called hysteresis. The loss factor, η, connected to the mechanical energy dissipation within the material, quantifies

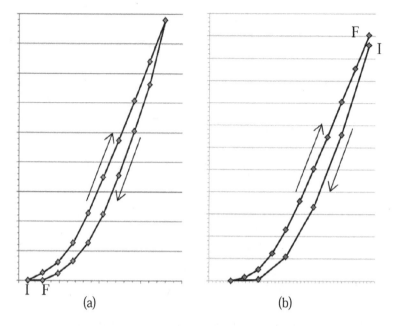

(a)   (b)

Figure A.21 Difference between a direct cycle (a) and an inverse cycle (b). In a direct cycle, starting from a beginning point I, there is a loading phase till a maximum point, which is followed by an unloading phase ending in correspondence to a final point F (the segment FI is only a connecting portion to close the cycle). An inverse cycle starts from point I (unloading phase) and ends in F (loading phase). Both a direct cycle and an inverse one are run clockwise, but in a direct cycle the starting point (I) and the arrival point (F) are in correspondence to loads close to zero, while an inverse cycle begins (I) with the fiber in traction and ends (F) with the fiber again in traction.

the mechanical (internal) damping associated with the analyzed specimen.

In the case of cycles coming from σ–ϵ curves (samples subjected to a tensile test with cyclical loading conditions), the loss factor, η, is defined as follows: it is the ratio (Fig. A.22) between the area of the hysteresis cycle (dissipated energy, D) and the subtended area of the loading curve of the hysteresis cycle (energy supplied to the fiber, U) multiplied by the factor $\frac{1}{2\pi}$; in mathematical terms:

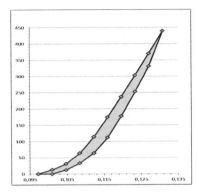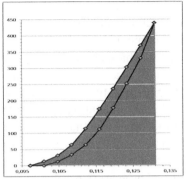

Figure A.22 Graphic example of the dissipated energy D in a cycle (namely the area of the cycle, on the left) and of the stored energy U into the cycle (on the right). The ratio between D and U multiplied by the factor $1/(2\pi)$ gives the loss factor, η.

$$\eta = \frac{D}{2\pi U}.$$

In other words the loss factor is a ratio among specific energies per radiant unit.

Thus the loss factor, η, expressed in percentage terms (and here denoted by the symbol $\eta_\%$) has this mathematical expression:

$$\eta = \frac{D}{2\pi U} 100.$$

We evaluated D and U areas by means of a computer by calculating the sum of many subtended trapezes.

Because we have defined a particular loss factor on the so-called inverse cycles (see below), we named the common percentage loss factor as "direct loss factor" and we utilized the special notation η_D.

(7) Inverse Cycle Loss Factor, η_I

In a loading cycle the plasticity can increase its area and so the loss factor.[17] We then decided to consider a different kind of loading

---

[17] This is, for example, the case of the formation of some fractures in the flax fiber's P layer (Section A.6). Since these fractures are irreversible, their contribution falls into the plastic component of the deformation.

cycle, which we called "inverse." While a direct cycle is a loading–unloading sequence (Fig. A.21a), an inverse loading cycle is obtained by coupling the unloading phase of the previous cycle with the loading phase of the following cycle (Fig. A.21b).

Since in a tensile test we can consider in general more than one inverse cycle, the inverse cycle loss factor, η_I, is evaluated referring to the last unloading phase of the previous cycle coupled with the last loading cycle that produced the fiber breaking.

This parameter is more significant than the direct percentage loss factor, $\eta_{D\%}$, if the loading cycles are not too numerous, as in the case of premature breaking of the tested fiber.

A.14 Cap. 9: List of the Plant Species from DNA Analysis

The following list shows the 24 different plant species that have been identified [58] by DNA analysis on the dusts of the Shroud; for each item we give the following details: species, family, common name, and origin.

(a) *Carpinus species*, Betulaceae, hornbeam, West and Central Asia, Eastern Europe and South America

(b) *Cichorium intybus*, Asteraceae, chicory, Africa, temperate Asia and tropical Europe

(c) *Cucumis sativus*, Cucurbitaceae, cucumber, tropical and temperate Asia

(d) *Equisetum species*, Equisetaceae, horsetail, temperate Asia, Australasia, Europe, and North America

(e) *Glycine max*, Fabaceae, soybean, South and tropical Asia

(f) *Humulus lupulus*, Cannabaceae, hop common, Africa, temperate Asia, Europe, and North America

(g) *Juglans species*, Juglandaceae, walnut, North America and South Eurasia

(h) *Lolium multiflorum*, Poaceae, ryegrass greater, Africa, temperate and tropical Asia, and Europe

(i) *Lolium species*, Poaceae, ryegrass, Africa, tropical and temperate Asia, and Europe

  (j) *Nicotiana species*, Solanaceae, tobacco, North and South America

  (k) *Picea abies*, Pinaceae, spruce, temperate and boreal regions of the northern hemisphere

  (l) *Picea species*, Pinaceae, spruce, temperate and boreal regions of the northern hemisphere

 (m) *Plantago argentea/lanceolata*, Plantaginaceae, plantain, Balkans and circumalpine areas

  (n) *Prunus salicina*, Rosaceae, plum, Africa, temperate Asia, and Europe

  (o) *Pyrus cossonii*, Rosaceae, pero, temperate and tropical Asia

  (p) *Pyrus pyrifolia*, Rosaceae, pero, temperate and tropical Asia

  (q) *Pyrus syriaca*, Rosaceae, pero, temperate and tropical Asia

  (r) *Robinia pseudoacacia*, Fabaceae, robinia, North America

  (s) *Salix species*, Salicaceae, willow Africa, temperate Asia and Europe

  (t) *Solanum species*, Solanaceae, eggplant wild, Africa and South America

  (u) *Trifolium fragiferum*, Fabaceae, strawberry clover, Africa, temperate and tropical Asia, and Europe

  (v) *Trifolium repens*, Fabaceae, white clover, Africa, temperate Asia, and tropical Europe.

 (w) *Trifolium species*, Fabaceae, clover, Africa, temperate Asia, and tropical Europe

  (x) *Vitis/Parthenocissus species*, Vitaceae, vine, Asia (Himalayan area)

Bibliography

1. AA, V. (2012). More from Fanti on the skewed nose (in Shroud of Turin Blog), URL http://shroudstory.com/2012/10/26/more-from-fanti-on-the-skewed-nose/#comments.

2. AA, V. (2014). Christianity in icons, murals and mosaic...), URL http://www.icon-art.info/hires.php?lng=en\&type=1\&id=3206\&mode=.

3. Accetta, J. S., and Baumgart, J. S. (1980). Infrared reflectance spectroscopy and thermographic investigations of the Shroud of Turin, *Applied Optics* **19**(12), pp. 1921–1929.

4. Adler, A. D. (1996). Updating recent studies on the Shroud of Turin, 625, pp. 223–228.

5. Adler, A. D. (2000). Chemical and physical characteristics of the blood stains, in *The Turin Shroud, Past, Present and Future. Proceedings of the International Scientific Symposium, Turin, March 2–5, 2000* (Effatà Editrice, Cantalupa, Italy), pp. 219–233.

6. Akin, D. E. (2010). Chemistry of plant fibres, in J. Mussig (ed.), *Industrial Applications of Natural Fibres. Structure, Properties and Technical Applications* (John Wiley & Sons), pp. 13–22.

7. Akin, D. E. (2013). Linen most useful: perspectives on structure, chemistry, and enzymes for retting flax, *ISRN Biotechnology* **2013**, pp. 1–23.

8. Andersons, J., Sparnins, E., Joffe, R., and Wallström, L. (2005). Strength distribution of elementary flax fibres, *Composites Science and Technology* **65**(3–4), pp. 693–702.

9. Arbelaiz, A., Cantero, G., Fernández, B., Mondragon, I., Gañán, P., and Kenny, J. (2005). Flax fiber surface modifications: effects on fiber physico mechanical and flax/polypropylene interface properties, *Polymer Composites* **26**(3), pp. 324–332.

10. Aslan, M., Chinga-Carrasco, G., Sorensen, B., and Madsen, B. (2011). Strength variability of single flax fibres, *Journal of Materials Science* **46**(19), pp. 6344–6354.

11. Aslanovski, D. (2012). academia.edu, URLhttp://independent. academia.edu/DavorAslanovski.

12. Aslanovski, D. (2012). Blog archive, URLhttp://deumvidere. blogspot.it/.

13. ASTM (1999). Standard specification for tensile testing machine for textiles, D76 99, ASTM.

14. ASTM (2007). Standard test method for tensile properties of single textile fibers, D3822 07, ASTM.

15. Baima Bollone, P. (1982). Indagini identificative su fili della Sindone, *Giornale della Accademia di Medicina di Torino* **1**(12), pp. 228–239.

16. Baima Bollone, P. (1998). *La Sindone la prova* (Mondadori, Italy).

17. Baima Bollone, P. (2000). *Sindone e Scienza all'inizio del Terzo Millennio* (Editrice La Stampa, Turin, Italy).

18. Baley, C. (2002). Analysis of the flax fibres tensile behaviour and analysis of the tensile stiffness increase, *Composites Part A: Applied Science and Manufacturing* **33**(7), pp. 939–948.

19. Baley, C., Busnel, F., Grohens, Y., and Sire, O. (2006). Influence of chemical treatments on surface properties and adhesion of flax fibre-polyester resin, *Composites Part A: Applied Science and Manufacturing* **37**(10), pp. 1626–1637.

20. Barbet, P. (1953). *A Doctor at Calvary: The Passion of Our Lord Jesus Christ as Described by a Surgeon* (P.J. Kenedy & Sons, New York).

21. Barbet, P. (1963). *A Doctor at Calvary* (Doubleday Image Book, New York).

22. Bellinger, A., and Grierson, P. (1966). *Catalogue of the Byzantine Coins in the Dumbarton Oaks Collection and in the Witthermore Collection*, Vol. 2, Part II (Dumbarton Oaks Centre of Byzantine Studies, Washington, USA).

23. Bevilacqua, M., Fanti, G., D'Arienzo, M., and De Caro, R. (2014). Do we really need new medical information about the Turin Shroud? *Injury Journal* **2**(45), pp. 460–464.

24. Bevilacqua, V., Ciccimarra, A., Leone, I., and Mastronardi, G. (2008). Automatic facial feature points detection, in H. De-Shuang, D. C. Wunsch, D. S. Levine and J. Kang-Hyun (eds.), *Advanced intelligent computing theories and applications. With aspects of artificial intelligence, Lecture Notes in Computer Science*, Vol. 5227 (Springer Berlin Heidelberg), pp. 1142–1149.

25. Bledzki, A. K., and Gassan, J. (1999). Composites reinforced with cellulose based fibres, *Progress in Polymer Science* **24**(2), pp. 221–274.

26. BPIM (2008). Guide to the expression of uncertainity in measurement, Tech. Rep., BPIM-JCGM 100, URL http://www.bipm.org/utils/common/documents/jcgm/JCGM_100_2008_E.pdf.

27. Breckenridge, J. (1959). *The Numismatic Iconography of Jiustinian II* (The American Numismatic Society, New York, USA).

28. Brillante, C. (1983). La fibrinolisi nella genesi delle impronte sindoniche, in *La Sindone, Scienza e Fede, Atti del II Convegno Nazionale di Sindonologia, Bologna 1981* (CLUEB, Bologna, Italy), pp. 239–241.

29. Brunati, E. (1993). Dobbiamo convincerci che il risultato del passaggio conclusivo del rapporto su nature è stato falsificato, *Collegamento pro Sindone*, May/June, pp. 50–56.

30. Brunati, E. (1997). A proposito di errori nel rapporto sulla datazione, *Collegamento pro Sindone*, May/June.

31. Brunati, E. (2005). Letter to Radicarbon 23/10/2005, Never published.

32. Bunsell, A. (2009). Tensile fatigue of textile fibres, in M. Miraftab (ed.), *Fatigue Failure of Textile Fibres* (Ed. Wood, Cambridge, UK), pp. 10–33.

33. Casarino, L., De Stefano, F., Mannucci, A., Zacà, S., Baima Bollone, P. and Canale, M. (1995). Ricerca dei polimorfismi del DNA sulla sindone e sul sudario di oviedo, *Sindon N.S.* **7**(8), pp. 36–47.

34. Cazzola, P., and Fusina, M. D. (1983). Tracce sindoniche nell'arte bizantino-russa, in *La Sindone, Scienza e Fede, Atti del II Convegno Nazionale di Sindonologia, Bologna 1981* (CLUEB, Bologna, Italy), pp. 129–135.

35. Charlet, K., Baley, C., Morvan, C., Jernot, J. P., Gomina, M., and Bréard, J. (2007). Characteristics of hermès flax fibres as a function of their location in the stem and properties of the derived unidirectional composites, *Composites Part A: Applied Science and Manufacturing* **38**(8), pp. 1912–1921.

36. Charlet, K., Jernot, J., Bréard, J. B., and Gomina, M. (2010). Scattering of morphological and mechanical properties of flax fibres, *Industrial Crops and Products* **32**(3), pp. 220–224.

37. Charlet, K., Jernot, J. P., Eve, S., Gomina, M., Bizet, L., and Bréard, J. (2010). Mechanical properties of flax fibers and of the derived unidirectional composites, *Journal of Composite Materials* **44**(24), pp. 2887–2896.

38. Charlet, K., Jernot, J. P., Eve, S., Gomina, M., and Bréard, J. B. (2010). Multi-scale morphological characterisation of flax: from the stem to the fibrils, *Carbohydrate Polymers* **82**(1), pp. 54–61.

39. Charlet, K., Jernot, J. P., Gomina, M., Bréard, J., Morvan, C., and Baley, C. (2009). Influence of an agatha flax fibre location in a stem on its mechanical, chemical and morphological properties, *Composites Science and Technology* **69**(9), pp. 1399–1403.

40. Coppini, L., and Cavazzuti, F. (2000). *Le Icone di CRISTO e la Sindone* (Ed. San Paolo, Cinisello Balsamo, Italy).

41. Cordero di San Quintino, G. (1845). *Delle monete dell'imperatore Giustiniano II. Ragionamento* (Stamperia Reale, Milan, Italy).

42. Crowfoot, G. M. (2012). Prodotti tessili, lavori di intereccio e stuoie, in C. Singer, E. J. Holmyard, R. S. Hall and T. I. Williams (eds.), *Storia della tecnologia*, Vol. 1/I (Bollati Boringhieri), pp. 420–454.

43. Csocsàn De Varallja, E. (1987). The Turin Shroud and Hungary, in G. Adriànyi, H. Glassl and E. Völkl (eds.), *Ungarn-Jahrbuch*, Vol. 15 (Dr. Rudolf Trfenik, München, Germany), pp. 1–49.

44. Damon, J. E., Donahue, D. J., Gore, B. H., Hatheway, A. L., Jull, A. J. T., Linick, T. W., Sercel, P. J., Toolin, L. J., Bronk, C. R., Hall, E. T., Hedges, R. E. M., Housley, R., Law, I. A., Perry, C., Bonani, G., Trumbore, S., Woelfli, W., Ambers, J. C., Bowman, S. G. E., Leese, M. N., and Tite, M. S. (1988). Radiocarbon dating of the turin shroud, *Nature* **337**(6208), pp. 611–615.

45. David, R. (1978). *Mysteries of the Mummies: The Story of the Manchester University investigation* (Littlehampton Book Services, London, UK).

46. De Liso, G. (2000). Verifica sperimentale della formazione di immagini su teli di lino trattati con aloe e mirra in concomitanza di terremoti, *Sindon N.S.* **14**, pp. 125–130.

47. Di Lazzaro, P., Fanti, G., Nichelatti, E., and Baldacchini, G. (2010). Deep ultraviolet radiation simulates the Turin Shroud image, *Journal of Imaging Science and Technology* **54**(4), pp. 1–6.

48. Enrie, G. (1933). *La Santa Sindone rivelata dalla fotografia* (S.E.I., Turin).

49. Fan, M. (2010). Characterization and performance of elementary hemp fibres: factors influencing tensile strength, *Bioresources* **5**(4), pp. 2307–2322.

50. Fanti, G. (2006). Sintesi della tesi sull'effetto corona e l'immagine sindonica, URL http://www.oratoriobeatopiergiorgiofrassati.it/oratorio/SacraSindone/TesinaSullaSindone-GuglielmoFanti.pdf.

51. Fanti, G. (2008). *La Sinfone, una sfida alla scienza moderna* (Ed. Aracne, Rome, Italy).

52. Fanti, G. (2010). Can a corona discharge explain the body image of the Turin Shroud? *Journal of Imaging Science and Technology* **54**(2), pp. 1–10.

53. Fanti, G., Angrilli, F., Baglioni, P., and Bianchini, G. (1991). Mechanical testing on a composite flexible structure, *Noordwijk: ESA*, pp. 375–381.

54. Fanti, G., Baraldi, P., Basso, R., and Tinti, A. (2013). Non-destructive dating of ancient flax textiles by means of vibrational spectroscopy, *Vibrational Spectroscopy* **67**, pp. 61–70.

55. Fanti, G., and Basso, R. (2008). Statistical analysis of dusts taken from different areas of the Turin Shroud, in *Shroud Science Group International Conference on the Shroud of Turin: Prospectives of a Multifaceted Enigma* (Columbus, USA), pp. 14–17, URL http://www.ohioshroudconference.com/papers/p16.pdf.

56. Fanti, G., Basso, R., and Bianchini, G. (2010). Turin Shroud: Compatibility between a digitized body image and a computerized anthropomorphous manikin, *Journal of Imaging Science and Technology* **54**(5), pp. 1–8.

57. Fanti, G., Calliari, I., and Canovaro, C. (2012). Analysis of microparticles vacuumed from the Turin Shroud, in *I International Congress on the Holy Shroud in Spain* (Valencia, Spain), pp. 1–25.

58. Fanti, G., and Gaeta, S. (2013). *Il mistero della Sindone* (Rizzoli, Milan, Italy).

59. Fanti, G., Lattarulo, F., and Scheuermann, O. (2005). Body image formation hypotheses based on corona discharge, in *Third Dallas International Conference on the Shroud of Turin* (Dallas, USA), pp. 1–28.

60. Fanti, G., and Maggiolo, R. (2004). The double superficiality of the frontal image of the Turin Shroud, *Jurnal of Optics A: Pure and Aplied Optics* **6**, pp. 491–503.

61. Fanti, G., and Malfi, P. (2014). Multi-parametric micro-mechanical dating of single fibers coming from ancient flax textiles, *Textile Research Journal* **84**(7), pp. 714–727.

62. Fanti, G., and Marinelli, E. (2001). A study of the front and back body enveloping based on 3d information, in *Dallas International Conference on the Shroud of Turin* (Dallas, USA), pp. 1–18.

63. Fanti, G., and Marinelli, E. (2003). *La Sindone Rinnovata: Misteri e Certezze* (Progetto Editoriale Mariano, Vigodarzere, Italy).

64. Fanti, G., Schwortz, B., and Accetta, A. (2005). Evidences for Testing Hypotheses about the Body Image Formation of the Turin Shroud, in *Third Dallas International Conference on the Shroud of Turin* (Dallas, USA), pp. 25–28, URL www.shroud.com/pdfs/doclist.pdf.

65. Faruk, O., Bledzki, A. K., Fink, H., and Sain, M. (2012). Biocomposites reinforced with natural fibers: 2000–2010, *Progress in Polymer Science* **37**(11), pp. 1552–1596, topical issue on polymeric biomaterials.

66. Fossati, L. (1977). Considerazioni sulle relazioni degli esperti che hanno esaminato la Sacra Sindone nel 1969 e nel 1973, in *Osservazioni alle perizie ufficialisulla Santa Sindone 1969-1976* (Centro Internazionale di Sindonologia, Turin, Italy), pp. 31–65.

67. Frache, G., Mari Rizzati, E., and Mari, E. (1976). Relazione conclusiva sulle indagini d'ordine ematologico praticate sul materiale prelevato dalla Sindone, in *La S. Sindone, ricerche e studi della commisisone di esperti nominata dall'Arcivescovo di Torino, Card. Michele Pellegrino, nel 1969* (Supplemento Rivista Diocesana Torinese, Turin, Italy), pp. 49–54.

68. Garello, E. (1984). *La Sindone e i Papi* (Ed. Corsi, Turin, Italy).

69. Garlaschelli, L. (2010). Lifesize reproduction of the Shroud of Turin and its image, *Journal of Imaging Science and Technology* **54**(4), pp. 1–14.

70. Garza-Valdes, L. (1999). *The DNA of God?* (Doubleday, New York, USA).

71. Ghiberti, G. (2002). *Sindone le immagini 2002* (ODPF, Turin, Italy).

72. Gilbert Jr., R., and Gilbert, M. M. (1980). Ultraviolet–visible reflectance and fluorescence spectra of the Shroud of Turin, *Applied Optics* **19**(12), pp. 1930–1936.

73. Gonella, L., Riggi di Numana, G., Pinna Berchet, G., Berbenni, G., Salatino, L., and Dubini, S. (2005). *Il giorno più lungo della Sindone. Cronache e documenti sulle operazioni di prelievo dei campioni per la radiodatazione del telo sindonico, 1986-1988* (Fondazione 3M Edizioni, Turin, Italy).

74. Guerreschi, A. (2002). Nuovi elementi rivelati dalla fotografia su due particolari: ferita al polso e occhio destro, in *Atti del Congresso Mondiale Sindone 2000* (Gerni Editori, Orvieto, Italy), pp. 41–46.

75. Guerreschi, A., and Salcito, M. (2002). Ricerche fotografiche e informatiche sulle bruciature e sugli aloni visibili sulla sindone e conseguenze sul piano storico, in *IV Symposium Scientifique International du CIELT* (CIELT, Paris), pp. 31–65.

76. Guerreschi, A., and Salcito, M. (2005). Further studies on the scorches and the watermarks, in *Third Dallas International Conference on the Shroud of Turin* (Dallas, USA), pp. 1–10, URL http://www.shroud.com/pdfs/aldo4.pdf.

77. Hawking, S. (1993). *Dal Big Bang ai Buchi Neri: Breve Storia del Tempo* (Supersaggi BUR, Milan, Italy).

78. Heller, J. H., and Adler, A. D. (1980). Blood on the Shroud of Turin, *Applied Optics* **19**(16), pp. 2742–2744.

79. Hu, W., Ton-That, M., Perrin-Sarazin, F., and Denault, J. (2010). An improved method for single fiber tensile test of natural fibers, *Polymer Engineering & Science* **50**(4), pp. 819–825.

80. Hughes, M. (2012). Defects in natural fibres: their origin, characteristics and implications for natural fibre-reinforced composites, *Journal of Materials Science* **47**(2), pp. 599–609.

81. Jackson, J. P. (1998). *Does the Shroud of Turin Show Us the Resurrection?* (Biblia y Fe).

82. Jones, S. E. (2014). Four proofs that the AD 1260–1390 radiocarbon date for the Shroud has to be wrong!: 2 the Vignon markings (2), Urlhttp://theshroudofturin.blogspot.it/2012/02/four-proofs-that-ad-1260-1390_16.html.

83. Judica Cordiglia, G. B. (1986). La sindone immagine elettrostatica? in *La Sindone, nuovi studi e ricerche, Atti del III Congresso Nazionale di Studi sulla Sindone* (Trani, Italy), pp. 313–327.

84. Judica Cordiglia, G. B. (1988). Ricerche ed indagini di laboratorio sulle fotografie eseguite nel 1969, in *Atti del IV Congresso Nazionale di Studi sulla Sindone, 1987* (Cinisello Balsamo, Italy), pp. 132–141.

85. Jumper, E. J., Adler, A. D., Jackson, J. P., Pellicori, S. F., Heller, J. H. and Druzik, J. R. (1984). A comprehensive examination of the various stains and images on the Shroud of Turin, in *Archaeological Chemistry III* (American Chemical Society), pp. 447–479.

86. K., C., P., J. J., M., G., C., B., L., B., and J., B. (2007). Morphology and mechanical behaviour of a natural composite: the flax fiber, in *16th International Conference on Composite Materials (ICCM-16), Kyoto, Japan*, pp. 1–8.

87. Kalia, S., Kaith, B. S., and Kaur, I. (2009). Pretreatments of natural fibers and their application as reinforcing material in polymer composites – a review, *Polymer Engineering & Science* **49**(7), pp. 1253–1272.

88. Kissinger, H. E. (1957). Reaction kinetics in differential thermal analysis, *Analytical Chemistry* **29**(11), pp. 1702–1706, URL http://pubs.acs.org/doi/pdf/10.1021/ac60131a045.

89. Kohlbeck, J., and Nitowski, E. L. (1986). New evidence may explain image on Shroud of Turin, *Biblical Archaeology Review* **12**(4), pp. 23–24.

90. Ku, H., Wang, H., Pattarachaiyakoop, N., and Trada, M. (2011). A review on the tensile properties of natural fiber reinforced polymer composites, *Composites Part B: Engineering* **42**(4), pp. 856–873.

91. Lattarulo, F. (1998). L'immagine sindonica spiegata attraverso un processo sismoelettrico, in *III Congresso Internazionale di Studi sulla Sindone* (Turin, Italy), pp. 334–346.

92. Lavoie, G. (2000). *Resurrected* (Thomas More, Texas, USA).

93. Libby, W. F. (1952). Radiocarbon dating, *Science* **116**, pp. 673–681.

94. Lilholt, H., and Lawther, J. M. (2000). Natural organic fibers, *Comprehensive Composite Materials* **1**, pp. 303–332.

95. Lindner, E. (2002). The Shroud of Jesus Christ: the scientific gospel to renew the faith in Resurrection, in *Atti del Congresso Mondiale Sindone 2000* (Gerni Editori, Orvieto, Italy), pp. 165–170.

96. Loconsole, M. (1999). *Sullte tracce della Sacra Sindone di Torino, un itinerario storico-esegetico* (Ladisa Editori, Bari, Italy).

97. Malantrucco, L. (1992). *L'equivoco Sindone* (LDC Leumann, Turin, Italy).

98. Malfi, P. (2012). *Design, Construction and Calibration of a Tensile Test Machine on Single Textile Fibers and Experimental Results, also for Dating Purpose*, Master's thesis, University of Padua, Italy.

99. Martin, N., Mouret, N., Davies, P., and Baley, C. (2013). Influence of the degree of retting of flax fibers on the tensile properties of single fibers and short fiber/polypropylene composites, *Industrial Crops and Products* **49**, pp. 755–767.

100. Martinschitz, K. J., Boesecke, P., Garvey, C. J., Gindl, W., and Keckes, J. (2008). Changes in microfibril angle in cyclically deformed dry coir fibers studied by in-situ synchrotron X-ray diffraction, *Journal of Materials Science* **43**(1), pp. 350–356.

101. Meacham, W. (2005). *The Rape of the Shroud: How the Christianity's Most Precious Relic Was Wrongly Condemned, and Violated* (LULU Press).

102. Moroni, M. (1983). L'ipotesi della Sindone quale modello delle raffigurazioni artistiche del Cristo Pantocrator. conferma numismatica, in *La Sindone scienza e fede* (CLEUB, Bologna, Italy), pp. 175–180.

103. Moroni, M. (1986). Teoria numismatica dell'itinerario sindonico, in *La Sindone, nuovi studi e ricerche* (Ed. Paoline, Cinisello Balsamo, Italy), pp. 103–122.

104. Moroni, M. (2000). L'iconografia di Cristo nelle monete bizantine, in *Le icone di Cristo e la Sindone* (Ed. San Paolo, Cinisello Balsamo, Italy), pp. 122–144.

105. Nechwatal, A., Mieck, K., and Reußmann, T. (2003). Developments in the characterization of natural fibre properties and in the use of natural fibres for composites, *Composites Science and Technology* **63**(9), pp. 1273–1279.

106. Nicoletti, A. (2011). *Dal Mandylion di Edessa alla Sindone di Torino* (Ed. dell'Orso, Alessandria, Italy).

107. Paleotto, A. (1598). *Esplicatione del Lenzuolo ove fu avvolto il Signore, et delle Piaghe in esso impresse, col suo pretioso Sangue confrontate con la Scrittura, Profeti e Padri con la notitia di molte Piaghe occulte, & numero de' Chiodi. Et con pie meditationi de' dolori. [Explication of the Sheet in which our Lord was enfolded, and of the Sores imprinted on it with His precious blood; then compared with the Bible, Prophets and Fathers, with the description of many hidden scourges and number of nails. And pious meditation of His sufferings]* (Rossi, Bologna, Italy).

108. Pallesen, B. E. (1996). The quality of combine-harvested fibre flax for industrials purposes depends on the degree of retting, *Industrial Crops and Products* **5**(1), pp. 65–78.

109. Pesce Delfino, V. (2000). *E l'uomo creò la Sindone* (Ed. Dedalo, Bari).

110. Petterson, R. (2012). Filatura e tessitura, in C. Singer, E. J. Holmyard, R. S. Hall and T. I. Williams (eds.), *Storia della tecnologia*, Vol. 2/I (Bollati Boringhieri), pp. 193–222.

111. Pfeiffer, H. (1986). *L'immagine di Cristo nell'arte* (Città Nova, Rome).

112. Phillips, T. J. (1989). Shroud irradiated with neutrons? *Nature* **337**(16), February, p. 594.

113. Picciocchi, P., and Picciocchi, P. (1979). L'impronta a epsilon sulla fronte dell'Uomo della Sindone: Nuova ipotesi sulla modalità di produzione, *Arch. di medicina legale, sociale e criminologia* **24**(3), pp. 261–274.

114. Placet, V., Trivaudey, F., Cisse, O., Gucheret-Retel, V., and Lamine Boubakar, M. (2012). Diameter dependence of the apparent tensile modulus of hemp fibres: a morphological, structural or ultrastructural effect? *Composites Part A: Applied Science and Manufacturing* **43**(2), pp. 275–287.

115. Porter, D. (2014). What are the Vignon markings? (in the Definitive Shroud of Turin FAQ), URL http://greatshroudofturinfaq.com/History/Greek-Byzantine/Pre-944AD/Vignon/index.html.

116. Power, B. A. (2002). An unexpected consequence of radiation theories for the Holy Shroud of Turin image formation: a possible repositioning of the burial linens in the tomb, in *Atti del Congresso Mondiale Sindone 2000* (Gerni Editori, Orvieto, Italy), pp. 183–194.

117. Quattrocchi, O. (2014). Il fuoco santo di pasqua, URLhttp://www.mariadinazareth.it/Luce%20Santa/Miracolo%20della%20Luce%20Santa.htm.

118. Raes, G. (1976). Rapport d'analyse, in *La S. Sindone, ricerche e studi della commissione di esperti nominata dall'Arcivescovo di Torino, Card. Michele Pellegrino, nel 1969* (Rivista diocesana Torinese, Turin), pp. 79–83.

119. Raes, G. (1995). Historique de l'échantillon, in *Le prélévement du 21/4/1988 – Études du Tissu, Actes du Symposium Scientifique International (Paris 1989)* (OEIL, Paris), pp. 71–74.

120. Riani, M., Atkinson, A. C., Fanti, G., and Crosilla, F. (2013). Regression analysis with partially labelled regressors: carbon dating of the Turin Shroud, *Statistics and Computing* **23**(4), pp. 551–561.

121. Riani, M., Fanti, G., Crosilla, F., and Atkinson, A. C. (2010). Statistica robusta e radiodatazione della Sindone, *Sis - Magazine marzo*, pp. 1–7.

122. Ricci, G. (1981). *The Holy Shroud* (Center for the Study of the Passion of Christ and the Holy Shroud).

123. Rinaudo, J. B. (1998). Image formation on the Shroud of Turin explained by a protonic model affecting radiocarbon dating, in *III Congresso internazionale di studi sulla Sindone* (Turin, Italy), pp. 474–483.

124. Rodante, S. (1994). *La Scienza convalida la Sindone* (Massimo, Milan, Italy).

125. Rogers, R. (2005). Studies on the radiocarbon sample from the Shroud of Turin, *Thermochimica Acta* **425**(1–2), pp. 189–194.

126. Rogers, R. N., and Arnoldi, A. (2002). Scientific method applied to the Shroud of Turin, a review, URL https://www.shroud.com/pdfs/rogers2.pdf.

127. Satyanarayana, K., Guimarães, J., and Wypych, F. (2007). Studies on lignocellulosic fibers of Brazil. Part I: source, production, morphology, properties and applications, *Composites Part A: Applied Science and Manufacturing* **38**(7), pp. 1694–1709.

128. Scheuermann, O. (2003). *Hypothesis: Electron Emission or Absorption as the Mechanism That Created the Image on the Shroud of Turin: Proof by Experiment*, 1st edn. (Fondazione 3M, Segrate, Italy).

129. Schwalbe, L. A., and Rogers, R. N. (1982). Physics and chemistry of the Shroud of Turin: a summary of the 1978 investigation, *Analytica Chimica Acta* **135**(1), pp. 3–49.

130. Sear, D. R., Bendall, S., and O'Hara, M. D. (1987). *Byzantine Coins and Their Values*, 2nd edn. (Seaby, Imprint of Spink & Son, London, UK).

131. Shah, D. U. (2013). Developing plant fibre composites for structural applications by optimising composite parameters: a critical review, *Journal of Materials Science* **48**(18), pp. 6083–6107.

132. Soyel, H., and Demirel, H. (2007). Facial expression recognition using 3D facial feature distances, in M. Kamel and A. Campilho (eds.), *Image Analysis and Recognition, Lecture Notes in Computer Science*, Vol. 4633 (Springer, Berlin, Heidelberg), pp. 831–838.

133. Stamboulis, A., Baillie, C., and Peijs, T. (2001). Effects of environmental conditions on mechanical and physical properties of flax fibers, *Composites Part A: Applied Science and Manufacturing* **32**(8), pp. 1105–1115.

134. Summerscales, J., Dissanayake, N., Virk, A., and Hall, W. (2010). A review of bast fibres and their composites. Part 1; fibres as reinforcements, *Composites Part A: Applied Science and Manufacturing* **41**(10), pp. 1329–1335.

135. Symington, M. C., Banks, W. M., West, O. D., and Pethrick, R. A. (2009). Tensile testing of cellulose based natural fibers for structural composite applications, *Journal of Composite Materials* **43**(9), pp. 1083–1108.

136. Thiruchitrambalam, M., Athijayamani, A., Sathiyamurthy, S., and A. Thaheer, A. S. (2010). A review on the natural fiber-reinforced polymer composites for the development of roselle fiber-reinforced polyester composite, *Journal of Natural Fibers* **7**(4), pp. 307–323.

137. Thomason, J., Carruthers, J., Kelly, J., and Johnson, G. (2011). Fibre cross-section determination and variability in sisal and flax and its effects on fibre performance characterisation, *Composites Science and Technology* **71**(7), pp. 1008–1015.

138. Tipler, F. J. (2007). *The Physics of Christianity* (Doubleday, New York, USA).

139. Torulf, N. (2006). Flax and hemp fibres and their composites: a letterature study of structure and mechanical properties, in N. Torulf

(ed.), *Micromechanical Modelling of Natural Fibres for Composite Materials* (KFS, Lund, Sweden), pp. 1–69.

140. Valtorta, M. (1986). *The Poem of the Man-God*, Vol. 1 (Nicandro Picozzi and Patrick McLaughlin, Portland, USA).

141. Valtorta, M. (2001). *L'Evangelo come mi è stato rivelato*, Vol. 10 (Centro Editoriale Valtortiano).

142. Van Dam, J. E. G., and Gorshkova, T. A. (2003). Plant growth and development: plant fiber formation, in D. J. Murphy, B. C. Murray, and B. Thomas (eds.), *Encyclopedia of Applied Plant Sciences, Ch. 46* (Elsevier, Amsterdam, Holland), pp. 87–96.

143. Van Haelst, R. (2000). The Shroud of Turin and the reliability of the 95% error confidence interval, in *Proceedings of the 1999 Shroud of Turin International Research Conference, Richmond, Virginia* (Magisterium Press, Glen Allen, Virginia, USA), pp. 321–325.

144. Vial, G. (1995). Le linceul de turin, étude technique, in *Le prélévement du 21/4/1988: Études du Tissu, Actes du Symposium Scientifique International (Paris 1989)* (OEIL, Paris), pp. 75–106.

145. Vignon, P. (1902). *The Shroud of Christ* (Ed. Dutton & Co., New York, USA).

146. Vignon, P. (1938). *Le Saint-Suaire de Turin devant la Science, l'Archéologie, l'Histoire, l'Iconographie, la Logique* (Masson et C. Editeurs, Paris).

147. Villarreal, R., Schwortz, B., and Benford, S. (2000). Analytical results on threads taken from the Raes sampling area (corner) of the Shroud, in G. Fanti (ed.), *Proceedings of Shroud Science Group International Conference: The Shroud of Turin; Prospectives on a Multifaceted Enigma, August 14–17, Ohio State University* (Libreria Progetto, Padua, Italy), pp. 1–2.

148. Volckringer, J. (1991). *The Holy Shroud: Science Confronts the Imprints* (The Runciman Press, Manly, Australia).

149. Wang, G., Yu, Y., Shi, S. Q., Wang, J., Cao, S., and Cheng, H. (2011). Microtension test method for measuring tensile properties of individual cellulosic fibers, *Wood and Fiber Science* **43**(3), pp. 251–261.

150. Whanger, A. (2000). Icone e Sindone, in *Le icone di Cristo e la Sindone* (Ed. San Paolo, Cinisello Balsamo, Italy), pp. 122–144.

151. Whanger, A., and Whanger, M. (1998). *The Shroud of Turin, an Adventure of Discovery Technique: A New Image Comparison Method and Its Applications* (Providence House, Franklin, USA).

152. Whanger, A. D., and Whanger, M. (1985). Polarized image overlay technique: a new image comparison method and its applications, *Applied Optics* **24**(6), pp. 766–772.

153. Whanger, A. D., and Whanger, M. V. (2007). The impact of the face image of the shroud on art, coins and religions in the early centuries, part 3, *Concil of Study of the Shroud of Turin, July 2007* **2**, 11.

154. Whilsom, I., and Schwortz, B. (2000). *The Turin Shroud: The Illustrated Evidence* (M O'Mara Books, London, UK).

155. Yu, Y., Jiang, Z., Fei, B., Wang, G., and Wangkun, H. (2011). An improved microtensile technique for mechanical characterization of short plant fibers: a case study on bamboo fibers, *Journal of Materials Science* **46**(3), pp. 739–746.

156. Zaninotto, G. (1988). Orazione di Gregorio il Referendario in occasione della traslazione a Costantinopoli dell'immagine edessena nell'anno 944, in *La Sindone, indagini scientifiche, Atti del IV Congresso Nazionale di Studi sulla Sindone, Siracusa, 1987* (Ed. Paoline, Cinisello Balsamo, Italy), pp. 344–352.

157. Zugibe, F. (2005). *The Crucifixion of Jesus, a Forensic Inquiry* (M. Evans & Co., New York).

Index

T - #0019 - 101024 - C20 - 234/156/24 [26] - CB - 9789814669122 - Gloss Lamination